NEW ENERGY TECHNOLOGY--
SOME FACTS AND ASSESSMENTS

The MIT Press
Cambridge, Massachusetts, and London, England

NEW ENERGY TECHNOLOGY--
SOME FACTS AND ASSESSMENTS

H. C. Hottel and J. B. Howard

Library of Congress Cataloging in Publication Data

Hottel, Hoyt Clarke, 1903-
  New energy technology.

  Includes bibliographical references.
  1. Power resources--U.S. I. Howard, Jack Benny,
1937- joint author. II. Title.
TJ23.H67   621.4   70-37654
ISBN 0-262-08052-4
ISBN 0-262-58019-5 (pbk)

CONTENTS

Contents

PUBLISHER'S NOTE

The aim of this format is to close the time gap between
the preparation of certain works and their publication in
book form. A large number of significant though special-
ized manuscripts make the transition to formal publication either after a considerable delay or not at all.
The time and expense of detailed text editing and composition in print may act to prevent publication or so to
delay it that currency of content is affected.

The text of this book has been photographed directly
from the authors' typescript. It is edited to a satisfactory level of completeness and comprehensibility
though not necessarily to the standard of consistency of
minor editorial detail present in typeset books issued
under our imprint.

The MIT Press

This volume is the product of several teaching and research activities. The first part of Chapter 1, including the figures, is an expansion of material prepared
for an energy workshop conducted by the authors in October 1970 as part of the Fifty Year Convocation of the
Chemical Engineering Department at M.I.T. Much of the
discussion of sulfur dioxide pollution and thermal pollution in Chapter 2 was developed from class notes prepared by one of us. The discussions of solar energy in
Chapters 6 and 7 are based on a number of years of association of one of us with M.I.T.'s Cabot Solar Energy
Project.

The largest and most important source of information
for the volume was a six-month study beginning February
1, 1971, of some of the energy problems of the United
States. The work was done under subcontract of the
Environmental Laboratory of M.I.T. to Resources for the
Future, Inc., in turn under contract to the National
Science Foundation to identify, within a comprehensive
framework, specific research needs related to the overall problem of satisfying national energy requirements.
Although the focus of attention of the M.I.T. group was
on research, the problems had to be put in reasonable
perspective. In the course of acquiring that perspective
and of transmitting research recommendations and the supporting background material to R.F.F., it became appar-
ent that there was considerable merit in the issuance of
the M.I.T. contribution as a separate volume. Presentation of the technical background for a research recommendation takes various forms dependent on whether the reader is a sponsor of the research, a researcher looking
for ideas on where significant work can be done, or a
lay reader wanting information. Our view has been that
research recommendations deserve to be assessed, and approved or challenged, and that this can occur only when

the process description is technically adequate. Issuance of this volume separate from the final R.F.F.-M.I.T.
report (scheduled for G.P.O. printing) frees the latter
from the need for recording the full technical background for some of the research recommendations.

In the course of the study discussions were held with
many experts, reports were read and compared, research
establishments were visited. The original plans for the
study were presented to the M. I. T. group by Messrs.
Sam H. Schurr and Hans H. Landsberg of R.F.F. The M.I.T.
group collaborated with Messrs. Schurr and Landsberg
throughout the study, but especially in the early stages
when the technical areas for assessment and the specific
program objectives were being identified. The authors
are grateful for the full-time and effective assistance
of Messrs. Steven Carhart and Richard Furman. Mr. Carhart
made a special study of energy transportation, prepared
all the material for that section and contributed to the
sections on oil shale, nuclear power and automotive power
plants. Mr. Furman made the computations on gas equilibrium, Figure 3-1, and made contributions to the Section
on fossil fuel conversion. Professor Peter Griffith prepared helpful material on automotive power plants and on
space heating. Thanks are due the following M. I. T.
faculty members who supplied briefings, criticisms, and
ideas: Professors Morris A. Adelman, Manson Benedict,
Michael C. Driscoll, Edwin R. Gilliland, Robert J. Hansen,
Myle J. Holley, Jr., Patrick M. Hurley, David D. Lanning,
Jean F. Louis, Edward A. Mason, Herman P. Meissner,
Norman C. Rasmussen, Philip Thullen, Neil E. Todreas,
David C. White, Glenn C. Williams, David G. Wilson.
Helpful discussions were held with Messrs. Harry Perry,
Neal Cochran, W. L. Crentz, Edw. Nicholson and L. Cooke,
with the members of the panel on Evaluation of Coal
Gasification Technology of the NAE-NRC Committee on Air

Quality Management, and with many individuals at the
U.S. Bureau of Mines Stations in Bruceton and Morgantown;
Bituminous Coal Research, Inc.; Institute of Gas Tech-
nology; Pope, Evans, and Robbins; United Aircraft
Research Laboratories.

The manuscript for this volume was completed in
early August, 1971.

Many subjects in the energy area generate controversy.
Although we tried to obtain views on both sides of questions about which there was much disagreement, the views
expressed and conclusions reached herein are our own;
for them we take full responsibility.

Hoyt C. Hottel
Jack B. Howard

Cambridge, Mass.
October 9, 1971

NEW ENERGY TECHNOLOGY--
SOME FACTS AND ASSESSMENTS

Chapter 1

INTRODUCTION AND SUMMARY

Although the magnitude of today's energy problems has
frequently been stressed, the importance of solutions
that are sound both socially and economically can hardly
be exaggerated. But the term "energy problems" has a
different connotation to different people. Some think
primarily of blackouts and the need for guaranteed continuity of electric power without visualizing how difficult
and, increasingly, how unacceptable is a continued growth
that follows the patterns of the past in meeting growing
demands. Others think primarily of environmental problems and are impatient with the slow change in the growth
patterns. Here the term "energy problems" relates to the
technological status of our energy and fuel conversion
processes, present and projected. Our objective will be
to determine and report this status, to assess the technical and economic adequacy of existing or proposed processes and their consistency with developing standards of
environmental quality, and to suggest where additional
effort--research, development, demonstration plant--is
needed to accelerate change.

In the study of a large problem, scope and depth are
always in conflict when manpower is limited. Although we
have attempted to be comprehensive, we have nevertheless
been unable in the allotted time to consider in depth all
of the many problems at issue, even by restricting our
interest in energy to its technology. As the work has
progressed, however, it has become clear that, just as we
have suffered from too broad a coverage to permit adequate inquiry on some issues, so have some past energy
studies suffered by narrowness of coverage. We thought
we had entered this study with reasonably open minds and
a fair view of the overall energy scene. As we read and
talked with others, we found we had unsuspected blind
spots, and we identified some mistaken prejudgments. Our

early views on the relative importance of several technical issues changed significantly during six months of in-
tensive study, and we came increasingly to realize how important a good view of the whole energy area is if one is
to judge the significance of action in a single area. If
several proposed solutions to a problem exist and only
one is presented by its inventor or sponsor to a disinterested party along with a request for assessment, disappearance of disinterest is almost normal: the assessor
tends to tell the sponsor what the latter wants to hear.

The manner of reporting a technological assessment
should depend on the audience, and when the interests of
the audience are varied and to a considerable extent unknown, there are problems. To a planner--say, a statesman or an economist--the conclusions from a technological
assessment suffice as inputs to his thinking; given the
statement that a certain process works, should cost about
so much to develop and so much to operate, and does not
violate environmental constraints, proper action can be
taken. To an engineer or scientist, on the other hand,
any process that is not in industrial use is almost by
definition replete with unsolved problems, and any stated
conclusions as to how much research or development is
necessary to make the process viable is subject to reval-
uation; an exposition of process details is an essential
background for acceptance of any stated conclusions. We
lean toward the second view but cannot claim consistency
of presentation of material. The extent to which process
detail precedes conclusions will vary greatly, sometimes
because we are addressing the nontechnical man, sometimes
because of the time and manpower limitations of this
study.

In this short chapter the record of past and present
United States energy production and use will be presented
briefly, followed by an even briefer projection of future

needs; the more important findings and assessments related
to research and development on energy use and energy conversion will then be summarized, by chapter.

1.1. Energy Supply and Demand, Past and Present

The annual United States consumption of wood, coal, crude
oil, natural gas, hydroenergy, and nuclear energy, all
expressed in quadrillions of Btu, is presented in Figure
1-1, which covers the period from 1850 to the present and
leaves room for extrapolation to the year 2000. Hydroenergy is converted to thermal units at the average thermal efficiency of fossil-fuel power plants operating in
the year in question (24.7% in 1955; 32% in 1968). Crude
oil includes natural-gas liquids. For conversion purposes the upper left corner of Figure 1-1 shows the Btu
equivalent of various fuel measures:

Quadrillion Btu	Amount of Fuel
58	10^{10} bbl crude oil
34.13	10^{13} kWh (thermal)
26	10^9 tons bituminous coal
19.5	10^9 cords 20th century wood
10.7	10^{13} cu ft natural gas
10.67	10^{12} kWh at 32% efficiency (1968 av.)

Figure 1-1 is full of history and drama, and readers with
a long memory will identify some of its irregularities
with vivid personal experience. Among the many points
meriting discussion are these:
a.
We enjoy, individually, such a short span of responsible
involvement in energy problems that we tend to lose track
of how similar, in some respects, the past patterns of
energy growth or replacement have been to those of our

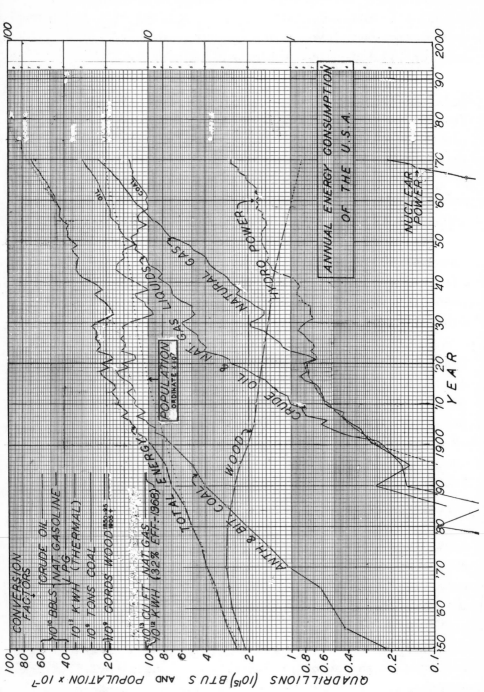

Figure 1-1. Annual Energy Consumption of the U.S.A.

day. We tend to think, because today's numbers are large,
that all changes of real significance have happened recently; and "recently" means "since I was about 25."
The logarithmic scale tends to cure this defect. The
most striking feature of the plot is the relentless and
almost constant upward march of energy consumption, which
causes the industrialist or engineer of 1971 to echo the
words of his great grandfather, "Never in history has
energy consumption matched ours."
b.
The past pattern of growth would not look as much like
the present one as it really does if wood, left out of
most modern compilations of energy, had been left out
here. Only when it is included is the line of growth
nearly straight back to 1850. Its inclusion in the present-day summation, however, adds only 1%.
c.
New fuels have always come in with a steep slope, not
generally displacing but adding to the contribution of
established fuels. Integration of a new fuel into the
economy has occurred in reasonably similar patterns three
times. By laying off distances along the edge of a piece
of paper corresponding to factors of 10 and 5 and sliding
the paper along under the top line to establish intersec-
tions with the various fuels, one finds the years at
which a fuel contributed 10% and 20% of the total. This
happened for coal in 1851 and sixteen years later, for
oil in 1918 and nine years later, for gas in 1935 and
seventeen years later. The average annual growth rate of
gas of 6% per year for 15 years from 1955 to 1970 is considered phenomenal; it was more than matched by petroleum
during the 22-year period 1934 to 1956.
d.
Coal has had a curious growth pattern. Its consumption
was 13.6 x 10^{15} Btu both in 1912 and in 1970; it has os-

cillated around that number, hitting highs of 17 x 10^{15}
in the war years 1918 and 1943 and lows of 9-10 x 10^{15} in
1932 and 1958. The trend has been upward for the last
9 years.

e.

Nuclear power is just coming into the picture; energy
from wood is estimated to be four times that from nuclear
sources. The latter is, of course, growing fast and is
somewhat comparable to petroleum in 1900.

The total energy consumption has been growing somewhat
faster than the population (dotted line), at least since
1885. This is better presented in Figure 1-2, which
shows the per capita energy consumption, millions of Btu
per year,* with a superimposed scale that expresses the
consumption as a multiple of the average adult caloric intake of 2800 kcal/day. Per capita consumption stayed constant for the first 35 years of the record, doubled between 1885 and 1920, fell precipitously during the great
depression, and has had a generally upward trend since
1932, ending in a steep rise since 1960. Consumption per
capita in 1970 was 338 x 10^6 Btu/yr, equivalent to about
6.7 gallons of petroleum per day or 80 times the human
caloric intake. This can be visualized as the equivalent
of 80 slaves working for each one of us to maintain our
modern affluent way of life. It is perhaps surprising to
realize that the last 50 years have seen less than a 70%
increase in per capita energy consumption.

Increasing energy consumption should produce an increase in goods and services. The former, measured by
the per capita gross national product (GNP)--expressed in
1958 dollars--appears as a second curve in the lower part
of Figure 1-2.

----
*The curve includes wood, which adds 10%, 5%, and 1% in
1912, 1940, and 1970.

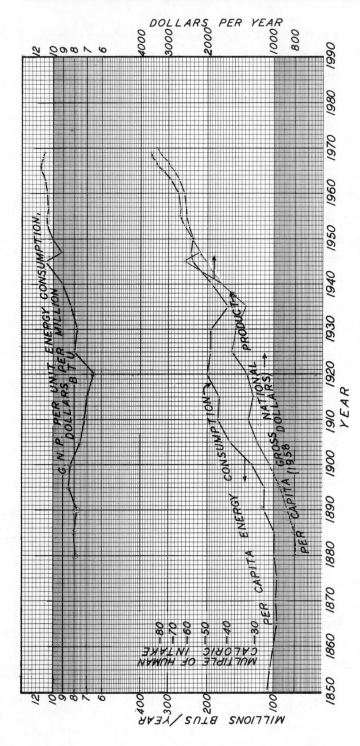

Figure 1-2. Per Capita Energy Consumption, Per Capita Gross National Product, and Their Ratio, as Functions of Time

If goods and services are produced in constant ratio,
the GNP per unit of energy consumed should measure efficiency of energy use. This ratio, $GNP/10$^6 Btu, appears
as the top line of the figure. The extent to which the
drop from 1900 to 1920 is a measure of decreased efficiency of energy use versus our limited ability to evaluate
old dollars versus a shift to more services relative to
goods is unknown. With some faltering the curve marched
upward from 1920 to 1955, indicating a continuing increase
in the effectiveness of using energy. From 1955 the
growth was slow, and since 1967 it has turned down. This
could mean that there has been a loss in efficiency of
use of energy; more probably it means that the ratio of
services to goods is increasing. Combined with the continuing increase in per capita energy consumption it is a
possible warning sign.

Figure 1-1 showed where our energy has been coming
from; Figure 1-3 shows how it is being used. Figure 1-3
is a double bar graph, divided vertically according to
category of use and horizontally according to fuel type.
In consequence, the relative contribution of a particular
fuel to a particular end use is measured quantitatively
by the area of the labeled item. For example, coal contributes 20.0% to the total energy input (bottom of left
column), and coal use by public utilities to generate
electricity is 56.7% of 0.20, or 11.3% of total United
States energy use. The sum of the separate fuel contributions in a single category of use appears in the small
bar graph on the right; that graph shows the division of
total fuel into household and commercial, industrial,
transportation, and electric power generation. (Hydro-
power is converted to equivalent fuel at the national-
average efficiency of conversion of fossil fuel to power,
32%.) As a general guide to placing problems in perspective, a handy approximation is as follows: about one-

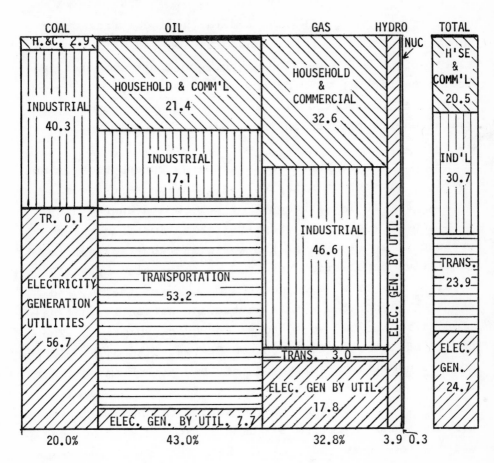

Figure 1-3. Distribution of U.S. Energy Consumption, 1970
(Preliminary Estimates by U.S. Department of the Interior)

third of our energy is consumed in the industrial category, one-fourth each in transportation and utilities
electricity, and one-fifth in household and commercial
use. The picture will of course change, particularly the
fractional consumption in producing electric power; electrical energy production is expected to double in 10
years.

The picture of United States energy consumption is not
complete without identifying its relation to the world
total. Figure 1-4 is a plot of per capita energy consumption of various geographical or national units--ex-
pressed as a fraction of the world average--against the
cumulative population expressed as a fraction of world
total.* The total area under such a graph is one, and
the area associated with a particular nation is that nation's fraction of world total energy consumption. The
chart indicates that the United States, with 5.76% of the
world's population (width of U.S. bar at bottom of chart)
consumed energy at 6.05 (or 5.7)* times the per capita
consumption of the rest of the world (the height of the
U.S. bar) and therefore consumed 34.8% (32.8%)* of the
world's energy in 1967. A striking feature of the graph
is its indication of how much additional energy would be
consumed if the rest of the world's appetite and capacity
to produce approached ours--there would be a sixfold increase. World consumption is in fact growing faster than
that of the United States, with West Germany, Japan, and

----
*The solid and dotted lines are based on two different
interpretations of data from United Nations--World Energy
Supplies (1970). The solid lines are from data in which
1 kWh of hydro, nuclear, or geothermal energy has the
value 3412 Btu; the dotted lines are from a Resources for
the Future analysis by Darmstadter and associates (1971),
in which 1 kWh electrical is given its fossil-fuel energy
equivalent at 32% conversion efficiency. Consumption
data from nations with much hydroelectric power are thereby raised.

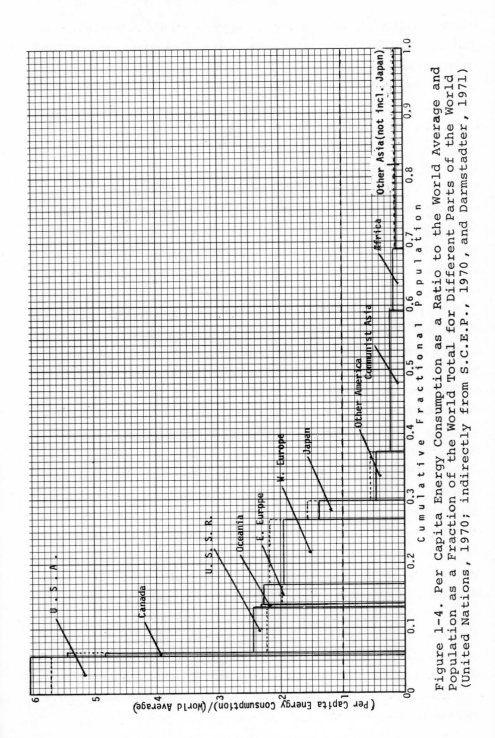

Figure 1-4. Per Capita Energy Consumption as a Ratio to the World Average and Population as a Fraction of the World Total for Different Parts of the World (United Nations, 1970; indirectly from S.C.E.P., 1970, and Darmstadter, 1971)

France leading the increase. According to Darmstadter
(1971), the energy consumption rate of every area or nation identified in Figure 1-4 is expected to grow faster
in the period 1965-1980 than that of the United States,
and the world fraction consumed by the United States is
expected to drop from 34.8 in 1967 to 26.8% in 1980.

1.2. Projection

Space, time, and the scope of this study prevent the inclusion here of an enormous body of additional statistical data on energy which are needed properly to consider
our energy problems of the future--data on costs, resources, discovery rates, and, particularly, projections
of need. If the discussion of Figure 1-1 suggested that
growth patterns of the future can be expected to follow
those of the past, that was not the intention. Growth
curves that maintain a constant percentage growth per
year (straight lines on Figure 1-1) cannot continue for-
ever in a finite world. This limitation has been clearly
understood by projectors of the past 40 years; in consequence, however, they have generally underpredicted future energy requirements. The top line of Figure 1-1 has
to bend, but when? There is increasing belief that the
answer is "soon" because of material and thermal pollution, but "soon" means 1985 to some, 2000 to others,
still later to others. Some of the many recent studies
of how to project the energy-demand curve have been summarized by Battelle (1969), with the conclusion that the
most probable growth rate from 1970 to 2000 is 3.2% per
year. An examination of Figure 1-1 indicates that 3.2%
(22-year doubling) is the exact average from 1931 or 1937
to 1970 (departure from that average is significant if
other intermediate years are chosen as the starting point).
The last nine years 1961-1970 represent the only part of
the curve since 1900 that is really straight; and the

average growth rate during that (phenomenal?) period has
been 4.57% (15.5-year doubling). Ritchings (1971) anti-
cipates a growth rate of 4.35% from 1970 to 1980 and 3.5%
from 1980 to 1990. These two projections probably represent the limits of present predictive judgment; their results are summarized in Table 1-1.

Regardless of which of these projections is accepted,
the conclusion is the now generally accepted one--that
energy demands of the end of this century will be enormous by present standards. If to this U.S. growth is
added the expected more rapid growth of the rest of the
world, it is clear that our capacity to handle the situation will require the effective contribution of the scientist, the engineer, the economist, the industrialist,
and the statesman.

Again, space and time prevent an adequate consideration of resources available to meet the above-projected
energy needs. Specific reference to resources is made in
the chapters on fossil-fuel conversion and nuclear energy.
Suffice it to say here that coal, shale oil, and tar
sands are available to supply our needs for centuries if
natural gas and oil run out; enough uranium is available
to satisfy our electrical needs for far longer when

Table 1-1. Probable Limiting Projections of U.S. Energy
            Requirements

Year	3.2% growth		4.35%, then 3.5%	
	10^{15} Btu	10^{12}kWh(t)	10^{15} Btu	10^{12}kWh(t)
1970*	69	20.2	69	20.2
1980	95	27.8	105	30.8
2000	177	51.9	210	61.5

*Tentative 1970 consumption: 68.8, excluding wood; 69.6
including it.

breeder technology is developed; and, if our conscience should bother us concerning our descendants in the year 3000, enough solar energy reaches the earth to supply our needs if we do not overpopulate the planet. But prodigal use of _any_ of our resources will be accompanied by an increasing fractional expenditure of our total productive effort on winning the energy needed for complementing that 1/25 horsepower device, the human body.

1.3. Summary and Conclusions

This section will abstract the more significant facts and assessments of this study, sometimes with additional comments. The organization will follow that of the chapters in sequence.

1.3.1. Background for Assessment of New Energy Technology

Common to the assessment of various fuel conversion processes is a body of technical information on energy transportation, energy storage, and pollution, covered in Chapter 2. Unlike the subject matter of later chapters, the material in this one is largely factual, with few comments on needs or implications of research or development. Energy Transportation. Guidance on the choice between moving fossil fuel to power-demand centers and transmitting electric power from a generating plant at the fuel source, avoidance of the high cost of solving the thermal pollution problems of a nuclear plant by power transmission from a thermally more favorable locale, moving coal to a gas plant near a center of gas demand versus long-distance piping of the gas from a cheap-coal area, balancing liquefied natural gas from Libya against locally-made synthetic gas--these are examples of need for comparative cost figures on energy transportation. Comparison on a common basis of cents per million Btu per hundred miles, (Figure 2-1), indicates the following:

a.

Electric power transmission in 500- to 700-kilovolt lines
is almost competitive with coal transport by unit train
after allowance for conversion efficiency.

b.

Coal pipelines or integral trains could cut the costs of
coal movement about 40%. It has to be remembered, however, that railroads' flexibility and hence ability to
institute new procedures are hindered by tradition, rate
regulation, labor problems, uncertainty caused by mergers,
etc.

c.

Coal pipelines or integral trains are almost competitive
with gas in moving energy if allowance is not made for
differences in efficiency of end use.

d.

The chief prospect for reduction in transmission cost of
gas lies in the development of lighter and stronger pipe
and improved pipe-laying techniques.

e.

Oil pipelines can transport energy at one-third the cost
of coal pipelines or integral trains or gas pipelines.

f.

For underground electric transmission the compressed-gas
insulated cable is expected to see increased use.

   Pipelines have increased in size, but the associated
scale economies in gas and oil transportation are so
great that the nation's projected rapid growth in energy
consumption will make still larger pipelines economically
sound.

SO_2 Pollution. It is clear that combustion-generated SO_2
may present a dispersion problem or a local pollution
problem, but it is not a global pollution problem; the

life of SO_2 is too short. It is estimated that by 1980
two-thirds of the total SO_2 emissions will originate from
fossil-fuel power plants. Contrary to widely held public
opinion, the conclusions of a year ago of a National Academy of Engineering Panel concerning SO_2 removal from
stack-gas still stands: the efficiencies of the various
processes for removal are not yet well established for
even the most advanced, and industrially proved technology for SO_2 removal does not exist. Present (July 1971)
estimates based to the extent available on pioneering utility experience with full-scale installations indicate
equipment costs that will add 0.6 to 1.2 mills/kWh to
power costs of new modern power stations with most of the
plants' economic life remaining, plus operating costs of
0.5 to 2.2 mills/kWh (more probably 1.0 to 1.5 mills).

These high figures make sulfur removal from the fuel
before combustion particularly attractive. Coal gasification under pressure with air, followed by gas scrubbing
to remove H_2S and use of the gas to operate a gas turbine
and steam turbine, looks particularly attractive, as does
coal liquefaction with minimum hydrogen use (see Section
1.3.2). The sulfur-from-power-plants problem cannot be
solved by use of naturally low-sulfur fuels; there aren't
enough at economic shipping distance from point of use.
Thermal Pollution. The 1970 world rate of energy consumption corresponds to an energy release rate 1/6000 that of
solar absorption by the earth. Thermal pollution is
therefore not a global problem and cannot be for over a
century. Local thermal water pollution, however, is a
problem, chiefly associated with electric power generation. Light-water-moderated nuclear power plants discharge about two-thirds more heat from the condenser into
cooling water than fossil-fuel plants because of their
lower thermal efficiency. Thermal pollution of water can

be prevented by using cooling ponds or wet or dry cooling towers, at added costs of power generation estimated at 0.1 mill, 0.1-0.2 mill, and 0.9-1.2 mills. The Federal Power Commission estimates an increase in waste-heat dissipation from power plants from 5.3×10^{15} Btu in 1970 to 28.4×10^{15} in 1990 (total energy consumption in 1970 was 69×10^{15} Btu). Since the projected evaporation rate for wet-tower and cooling-pond operation will then be 0.8% of the nation's runoff, extensive micrometeorological studies are warranted. Thermal water pollution at power plants can be eliminated by use of gas turbines or reduced by use of the gas-steam cycle.

Energy Storage. In the field of storage battery development, marked superiority over lead-acid batteries has been achieved, in nickel-cadmium batteries, for example; but lead-acid continues to be better economically. High-temperature systems being developed--sodium-sulfur and lithium-chlorine, for example--are superior to lead-acid and claimed to be economically competitive; opinion is divided on whether the obvious containment problems associated with safety will be solved to the satisfaction of the public.

For power-plant operation pumped-water storage is beginning to be used, with 10,600 MW planned as of December 1969.

Thermal storage for flattening the load curve of electric space heating or for solar house design is best achieved with water tanks, crushed gravel bins, or concrete blocks (used in England); no good heat-of-fusion material that operates near room temperature has been found.* Magnetite (magnetic iron ore) has an extraordinarily high volume heat capacity; a cylinder 2 feet in diameter and 12 feet long raised 600 F would store one day's space-heating requirements of a small (1200 gal fuel oil/

---

*Thickened and nucleated Glauber's salt is a possibility.

season) house.

1.3.2 Fossil Fuel-to-Fuel Conversion

More time went into the study of the important subject of
fuel-to-fuel conversion processes than into any other, and
the chapter covering it occupies about a third of this
volume. The sections cover pipeline-quality gas from
coal, low-Btu clean gas from coal, oil from coal, tar
sands and oil shale, and a critical comparison of relative merits of the various processes.

Pipeline-Quality-Gas. Following a brief section on gasification principles, eight processes are described, with
flowsheets, for making pipeline-quality gas from coal.
Economic assessments of probable cost are presented where
available. A comparison of the technical features of the
various processes, including advantages, disadvantages,
and questionable areas, is then made in condensed tabular
form (Table 3-1); and this is followed by a presentation
of research and development needs.

There is strong evidence, not presented in this volume,
that the 6% per year growth rate in natural gas use
(see Figure 1-1) will combine with continuously decreasing discovery rates of new gas to produce a great short-
age of natural gas if both trends persist. The United
States will need gas from coal at a time not established,
but very probably soon, on a scale that will dwarf all
other industry of comparable chemical content. In consequence, the Office of Coal Research, the Bureau of Mines,
and the American Gas Association are vigorously pursuing
research on synthetic gas from coal. Four processes--two
in pilot stage--deserve strong support; they have demonstrated sufficiently promising results, freedom from serious faults, and differences in concept to warrant being
kept in the running during the period of pilot-plant assessment of the processes. The four organizations in-

volved are the Institute of Gas Technology (Hygas in its three variations), Consolidation Coal Co. (CO_2 Acceptor), U.S. Bureau of Mines (Synthane and Hydrogasification), and Bituminous Coal Research (Bigas). The first two of these have pilot plants for the main gasification process near completion; the last two have plans for pilot plants and have done the necessary research. In view of the magnitude of the final commercial operation based on the research of these four teams to date, the nation cannot afford to overlook the possibility that any one of the four processes may have advantages properly assessable only on a larger scale than that of past laboratory research. The analogous case in nuclear power plant development has been the retention, at great cost during the development stages, of at least five varieties of slow reactors and four varieties of breeder reactors, simply because the time lost in going back to a complex process abandoned in an early stage is too great. Viewed another way, a retention of four processes through sufficiently advanced stages to permit making a final choice of process that saves 10¢/1000 cu ft in manufacturing costs will, in the days when our consumption of synthetic pipeline gas is one-third that of natural gas in 1970, amount to a saving of $750 million a year. We cannot afford to decide prematurely on the wrong process.

The strong recommendation of continuation of four processes at least through the pilot-plant stage implies that much basic research has already been carried out on coal gasification, but it does not imply absence of the need for more. Any process that wins the race will be in need of continuous modification and improvement; and a strong research program, both basic and applied, on the physics and chemistry of coal gasification is needed. Suggestions concerning that program are made in Section 3.1.7., subsections entitled "Gasification Fundamentals,"

"Entrained Flow," and so forth.
Low-Btu Clean Gas from Coal. The air-blown gas producer
that left the American scene in the twenties is due for a
comeback if it can be made larger, automatic, and capable
of handling caking coals, and if it can be given a high
capacity through operation under pressure. A major impetus for developing a process for making low-Btu gas from
coal comes from the electric power industry. Producer
gas made under pressure and scrubbed to remove sulfur is
an ideal gas-turbine fuel; and the combining of gas producers with gas-steam power cycles of high efficiency
looks attractive as a solution to the sulfur problem in
power production.

First-generation low-Btu clean gas producers using
countercurrent flow of lump coal and air can be purchased
today from Lurgi. Gasification and cleaning costs (with
no credit taken for recovered sulfur) are estimated to be
31.7¢ per million Btu in the product gas, which has a
higher heating value of 173 Btu/SCF.

Second-generation processes meeting the requirement of
90 g sulfur/10^6 Btu could use one of several gasification
schemes, all based on gasification of concurrently flowing air, steam, and pulverized coal. The Texaco Partial
Oxidation Process with moving pebble-bed heat recovery
system and hot carbonate scrubbing of H_2S is illustrative
of second-generation processes. Gasification and cleaning costs are estimated to be 17.6¢ per million Btu in
the product gas, which has a higher heating value of
179 Btu/SCF.

The implications of clean low-Btu gasification to power production, discussed in Section 1.3.4., are so striking as to warrant a more vigorous support of air gasification of coal at moderate pressures than now exists.
The Bureau of Mines' stirrer development for handling
highly-caking coals in Lurgi-type fixed-bed producers de-

serves strong support and should be expanded to include
solution of the dust-soot-tar problem. Entrained-flow
gasification has progressed far enough to justify pilot-
plant operation, independent of and without waiting for
progress in gasification with oxygen to make pipeline-
quality gas. On the level of basic research on pressure
gasification of coal with air, Section 3.2.3. outlines
some needs.

Oil from Coal. Failure to couple the expectation that
nuclear fuel will in the long run supply our power with
the recognition that this cannot happen fast enough to
eliminate the need for clean fossil fuels now is a real
national danger; inadequate emphasis on research to produce clean fossil fuels both for power and for processing
operations could produce an impasse between industrial
energy needs and the need to protect the environment.
Among the processes considered in Section 3.3 on "Oil
from Coal," Solvent-Refined Coal, the Consol Process for
Synthetic Crude Production, H-Coal, and COED all have merit and different objectives.

Solvent-refining of coal was initiated with the limited objective of producing a low-cost antipollution alternative to residual oil and natural gas for use under
boilers. Using minimal hydrogen--30 to 40 pounds per ton
of coal--and a pressure of 1200 psi, the process puts 90%
of the coal carbon into solution at an estimated cost,
exclusive of fuel, of 3 to 19¢ per million Btu, depending
on whether coke is made as a by-product. The process appears to deserve support through the pilot-plant stage.

The Consol process has been in development for some
years, went into pilot-plant operation in 1967 to make
liquids in the gasoline range, and was christened Project
Gasoline. Later studies indicated the advisability of
changing the objective to the manufacture of low-sulfur
synthetic crude oil, with extensive modifications of the

pilot plant being required. It is clear that until progress has been made on the proposed pilot-plant operation,
no real assessment of the prospects of the process can be
made. Cost estimates in 1969 indicated the possibility
of manufacturing synthetic crude at $3.25 per barrel from
Western coal at $1.25/ton, with an increase of 35¢ per
barrel for every $1 increase in coal cost. Crude prices
in Wyoming in 1969 were $3.75/barrel.

The H-Coal process uses a cobalt molybdate catalyst to
hydrogenate coal at 2700 psi in an ebullating bed. The
process has had bench-scale development, and research on
it at $1 million per year for 1971 and 1972 is to be sup-
ported jointly by five oil companies; plans for a prototype plant are expected in mid-1972. An economic projection of expected profits from a plant of 100,000 bbl/day
capacity indicated a total cash flow return on the investment of 15.2% to 18% for Illinois coal and 14.3% to 16.4%
for Wyoming coal.

The Char Oil Energy Development Process (COED) is
based on a multistage fluidized-bed pyrolysis of coal to
produce oil, gas, and char, followed by hydrotreating of
the oil to yield a petroleum refinery feedstock. The concept is to minimize the loss of hydrocarbons that occurs
when cracking is too severe by arranging a staged increase in temperature. The process yields only 1.2 barrels of oil per ton of coal, compared with about 3 barrels from Consol or H-Coal. Its success consequently depends on the marketing of the large char and gas by-product streams or on the use of the last to make hydrogen.
Economic studies of the process predate the acquisition
of data on the pilot plant; they are very favorable, indicating a discounted-cash-flow (DCF) rate of return of
16% when the product gas is used to make hydrogen and a
part of the char is used for process heat.

The most pressing research and development needs in

processing coal to make oil are funds to implement a
National Academy of Engineering recommendation concerning
operation of the Consol process pilot plant and funds to
construct and operate a solvent extraction plant for making power-plant fuel. Associated bench-scale research
problems are listed in Section 3.3.8.

Comments on the position of oil-from-coal processes in
relation to other conversion processes will be withheld
for the end of this section.

Oil from Tar Sands and Oil Shale. The need for technological sophistication in recovering oil from nonpetroleum
fossil deposits is much less when the raw material is tar
sands or oil shale than when it is coal. This may cause
earlier developments of oil from shale than from coal,
regardless of ultimate comparative economics.

Tar sand treatment has become commercial in Alberta,
on sands averaging 12.5% bitumen. The estimated return
on the investment (DCF) came to 5.8% when the value of
crude oil was taken as $2.90/bbl (equivalent to $3.60 in
Chicago) and a reasonable estimated plant cost was used
to replace the high published cost of a first plant.

Oil shale is more attractive than tar sands to the U.S.
oil industry largely because most of the good oil shale
is in Colorado rather than in Alberta, but partly because
the hydrogen-carbon ratio is higher. Three retorting processes have operated on sufficient scale (260-1000 tons/
day) to evaluate the retorting operation, and cost estimates of a 50,000 bbl/std day plant indicate a 9.9% rate
of return on the investment (DCF) when the syncrude price
is set at $3.20/bbl (equivalent to $3.60 in Chicago).
Experience, however, with mining and acceptably disposing
of shale at a rate in excess of one ton per second is
missing. German brown-coal strip-mining practice deserves imitation. The status of development and research
needs are discussed in Section 3.4.2.

Comparison of Proposed Fossil Fuel-to-Fuel Conversions.
An intercomparison of the conversion processes considered
--pipeline-quality gas, low-Btu gas from coal, oil from
coal by solvent refining, hydrotreating and staged pyrolysis, oil from tar sands and oil from shale--is difficult
because of different objectives of different customers,
national energy policy, foreign energy policy and import-
pricing, differences in status of development of the processes, the ability to predict resources, the differences
in cost-estimating procedures used in the economic
studies, and the degree of interchangeability of the products. These points are elaborated more fully in Section
3.5.1. Within the limitations they impose, a comparison
within three groups is logical: clean power production;
oil from the three major nonpetroleum sources; a three-
way comparison of pipeline-quality gas from coal, oil
from any source, and, sometimes, low-Btu gas locally made.
Clean Power from Coal. The shortcomings of stack-gas SO_2
cleanup systems so far investigated make processes for
production of sulfur-free fuels or for sulfur removal during combustion look especially attractive. The second of
these has not been properly examined here because of time
limitations. The remaining alternatives are as follows:
a.
Use of low-sulfur steam coals. Supply limitations make
this a short-time solution, at a premium on coal of about
$2/ton (ca. 0.8 mill/kWh). Coal cleaning would be capa-
ble of lowering the sulfur content to 1% in only about
one-quarter of the U.S. steam coal.
b.
Use of low-sulfur Western coals. Large supplies of Western coals and lignites containing less than 1% sulfur are
available, but most power-plant furnaces have not been
designed for handling such fuels. More important, the
cost of transportation 1500 miles by unit train adds

about $9/ton to base cost. Development of integral
trains might halve this number.
c.
Use of low-sulfur oil. Only about 13% of U.S. central-
station power was based on oil in 1970 (up markedly from
earlier years). Most U.S. oil is high in sulfur, and removal adds 50-80¢/bbl to the cost (0.8-1.3 mills/kWh).
The oil industry is investing heavily in sulfur-removal
plants, but residual oils are expensive to treat. The
U.S. oil consumption rate presently exceeds the finding
rate, and it is improbable that fossil fuel's contribution to the anticipated doubling of power generation in
ten years will come from oil.
d.
Reliance on present coal supplies, combined with stack-
gas cleanup. This has been found a more expensive operation than was claimed a few years ago, and stack-gas
treatment is expected to add $20 to $40/kW to power-plant
capital costs and 1 to 1.5 mills/kWh to operating expenses. The higher of the capital costs would add 0.8
mill/kWh to the power cost in a new installation, and
much more to that of a plant with less than half its useful life remaining.
e.
Use of low-Btu clean gas from coal, in the gas-steam cycle. In Section 5.2 the combination of a pressurized gas
producer with an advanced-cycle gas turbine followed by a
waste-heat boiler and steam turbine is shown to be feasible now (German and Russian practice). A cost analysis
believed to be realistic indicates that a present-generation clean gas-steam plant (construction in early 1970s)
can produce power at only 1 mill more than conventional
use of coal; to the latter about 2 mills should be added
to cover SO_2 stack-gas treatment. The second-generation
gas-steam plant was estimated to produce power cleanly by

the 1980s for 1 mill _less_ than a conventional coal plant
before SO_2 scrubbing is added. Thermal water pollution
from such a plant is very low.
f.
Use of ash-free, sulfur-free coal extract. It is pointed
out in Section 3.3.3. that, if the objective in treating
coal is not to make simulated crude oil but to cause only
sufficient liquefaction to permit elimination of sulfur
and ash, the expensive hydrogen needed for liquefaction is
minimized. The product may be burned hot, in a power
plant close-coupled to the refinery plant, by atomization
in a manner completely analogous to smokeless burning of
tar and pitch, or it may be shipped cold in heavy-flake
form and burned as pulverized fuel. It is claimed that
the clean ash-free fuel can be sold for between 3.3¢ and
13.3¢/10^6 Btu above the fuel cost. Although this would
add 0.3 to 1.3 mills/kWh to power costs, the credit due
to absence of need for precipitators and ash handling
amounts to 0.36 mill/kWh. Thus, even at the highest operating cost, labeled definitely off-optimum, the operation is projected to cost less than scrubbing SO_2 from
stack gases.

The above six alternatives indicate clearly that two
concepts, making clean low-Btu gas from coal and making
an ash-free, sulfur-free heavy hydrocarbon from fuel,
should be of great interest. If the cost estimates are
realistic--and they appear to be--completion of development of both processes should be supported. This calls,
in the first case, for federal funding of the coal extraction pilot plant that has been designed and, in the second, both for a more vigorous support of the work on coal
gasification with air and for the development of larger
advanced-cycle gas turbines.

Comparison of Coal, Tar Sands, and Oil Shale as Oil
Sources. All three of these sources of oil are so enormous that they must certainly someday be tapped. The
timetable for this, however, is very difficult to establish, especially because of changes in import quotas and
in prices established by foreign governments.

On a comparative basis tar sands appear not to have as
good a chance of early development as oil shale. Comparable past analyses yielded projected rates of return of
5.0% and 9.9% (discounted cash-flow method) for tar sands
and oil shale, respectively. In addition to this margin
of shale over sands, which probably exceeds the error in
calculations but which could disappear quickly as a result of a technological or a political development, oil
shale has the advantage that the largest North American
deposits are in the Green River area (mostly Colorado)
whereas the tar sands are mostly in Alberta. It appears
now that the processing of oil shale in the United States
is warranted earlier than extensive processing of tar
sands, barring further significant relative technological
change in the two processes or the imposing of Western
state restrictions on the mining of oil shale.

With respect to coal versus oil shale as sources of
synthetic crude oil the assessment is even more difficult.
Despite the considerable effort that has gone into oil-
from-coal research, processes for oil shale treatment are
simpler and more nearly ready for use. But the yield
from coal is so much greater (3 to 3.5 barrels per ton
versus 0.8 barrel for relatively rich oil shale) and the
disposal problem so much simpler that vigorous pursuance
of pilot-plant development followed by demonstration-
plant operation is in the best national interest. The
very fact that the processing of coal to produce oil is
considerably more complex than the production of oil from

shale is reason to expect a greater improvement, through
research, in the efficiency and cost of the process.

The development of any of the previously discussed oil
recovery processes to the point of producing synthetic
crude at prices reasonably near present crude prices
could well turn out to be primarily an insurance policy
but a very valuable one--insurance against excessive increase in the price of imported oil. On this ground
alone strong federal support of fuel conversion research
is eminently warranted.

Gas from Coal versus Synthetic Oil from Coal. A technical affirmation popular today among engineers knowledgeable in the energy field is this: "We need to learn how
to make both high-Btu gas and oil from coal, but gas from
coal comes first because we are running out of natural
gas." Without denying the validity of the conclusion, the
next few paragraphs will briefly examine the bases for it.

Data on annual oil and natural gas consumption, when
plotted on a logarithmic scale versus time, produce reasonably straight lines of markedly different slope. The
1960-1970 growth rates were 3.9%/yr for oil and 5.76%/yr
for gas. If these two growth rates are not to change,
the curves would cross in 1986 at 55×10^{15} Btu per year,
corresponding to 55 trillion cu ft gas and 9.5 billion
barrels of oil per year. No one is guilty of quite so
naive a prediction; the growth rates are generally assumed to decrease. But conventional projections of demand nonetheless tend to be extrapolations from the past
into the future unmodified by considerations of cost and
of interchangeability of energy types, coupled with a projection of production of the natural products, gas and
oil. Paired subtractions then yield the synthetic gas and
oil needs of the country. These needs, arrived at substantially without a consideration of anticipated costs
of making synthetic oil or gas, are sometimes presented

as a basis for decisions on needed research.

If natural gas and oil were available to a consumer
and the forces of the marketplace were allowed to act
freely, uncluttered by any institutional underbrush,*
what relative prices could a Btu bring in the two fuel
forms; for what use is oil at x% of the price of natural
gas a bargain? The reason for asking the question in the
present context is that if bench-scale or small pilot-
scale research suggest that synthetic oil can be made--
from anything--at y times the cost of synthetic gas, and
if y is less than x (and the market is large), then research on completion of development of the subject process should be pursued vigorously. The tradeoff value of
Btu's in oil versus gas varies, of course, with the character of use; there are many operations for which an x of
80% or even 90% would warrant use of the cheaper fuel.
Data from Chapter 3 will now be used to see what is known
or suspected about y.

Estimated prices of pipeline-quality synthetic gas at
the point of production were 55¢-68¢/10^6 Btu (equivalent
to \$3.19-3.95/bbl oil) from coal at \$4.60/ton, and 47¢/10^6
Btu (\$2.72/bbl oil) from lignite at \$2.25/ton. But these
figures were labeled optimistic, and the following were
offered as more probable:

Gas from coal, 85¢ and \$1 (\$4.93 and \$5.80/bbl);

Gas from lignite, not given, but subject to similar escalation.

These very "iffy" numbers may be compared with those from
Section 3.3 on oil from coal:

Consol, \$3.25/bbl for S-free syncrude from Western coal
at \$1.25/ton; \$3.58/bbl for S-free syncrude from Western
coal at \$2.25/ton;

----
*The name for public utility price regulation, states'
rights, import quotas, and other artifacts of our political structure.

H-Coal, $3.78/bbl for S-free furnace oil from Illinois
No. 6 coal at $3.25/ton; DCF rate of return = 10%;

COED, $4.00/bbl for S-free high-naphtha high-aromatic
syncrude from coal at $3.00/ton; DCF rate of return = 13%,
10%, or 16%, depending on mode of operation;

and with those from Section 3.4 on S-free oil from sand
and shale:

Tar sands, $3.60/bbl in Chicago; $2.90 in Alberta; DCF
rate of return = 5.8%;

Oil shale, $3.60/bbl in Chicago; $3.20 in Colorado; DCF
rate of return = 9.9%.

That the oil prices which were used in making these analyses are reasonably up to date is indicated by the fact
that July 1971 crude prices quoted in the Oil and Gas
Journal can all be expressed by the relation

Price, $/bbl = ($3.25 to $3.48) + 0.02(API Gravity - 30)

The spread reflects primarily the effect of crude sour-
ness.

   Comparison of the estimates of synthetic pipeline gas
costs with syncrude costs is especially difficult because
of the different accounting procedures used. Gas prices
regulated by the FPC are obtained by allowing 7% rate of
return and finding the synthetic gas cost, which is of
the order of twice* the present cost of natural gas. Private industry syncrude prices are assumed the same as the
market prices on crude oil, and the rate of return from
the operation is established. On the assumptions that
sulfur-free syncrude can be made by at least one process
for $3.60 with a 7% rate of return and that the best esti-
mate of synthetic pipeline gas price--in oil-equivalent--
is $5/bbl, one arrives at the conclusion that energy in
----
*This is an oversimplification; natural gas is bought, for
shipment through pipelines, at 22¢ to 60¢/1000 cu ft.

synthetic oil form will cost 72% as much as in pipeline
gas form, before transportation. On the average, oil
costs about half as much to transport by pipeline as gas
(Figure 2-1); expressed as a price differential, the figure is 50¢ per barrel per 1000 miles.

It is clear that no claim can be made that the above
numerical comparison is known to be valid. A well-
grounded comparison will not be possible until pilot
plants for gas and for oil have led to demonstration
plants and the latter have been operated long enough to
have reached near-optimization. If we consider the billions of dollars per year hinging on the outcome of such
a comparison, the need for large federal expenditures to
develop clean synthetic fuels seems obvious.

It has not been the intention to imply in the above
presentation that the cheaper fuel will win the race.
Gas has advantages over oil which will offset a price difference (opposite in sign to the one existing today). As
an aid to making more quantitative any comparison of gas
with oil there is need for a comprehensive study of U.S.
industry to determine the degree of interchangeability of
fuel feasible at various cost differentials.

Related to the argument of the last several pages is
the consideration of pipeline-quality gas versus low-Btu
clean gas, the latter either made locally by the plant
needing it or possibly considered for distribution from a
central plant. Presently-available Lurgi pressure gasifiers can make clean producer gas for $57.7¢/10^6$ Btu.
Since pipeline gas processes have not yet been developed,
it is fairer to compare pipeline gas with what can be
achieved in second-generation producers; the claimed cost
is $52.1¢/10^6$ Btu. This cost takes into account the purchase of coal at 70% above the cost to a pipeline plant
because of absence of mine ownership. It thus appears
that clean producer gas might be able to supply industri-

al plants locally with gaseous fuel at a considerable
saving over synthetic pipeline gas. This problem justifies a study similar to that on oil versus gas, to determine what markets would in the long run be better served
by producer gas than by pipeline gas.

1.3.3. Nuclear Power

The technological and commercial success of nuclear reactors for power production is attested by the fact that at
the end of 1970 a total of 20 operable nuclear plants contributed 2.2% of the nation's electric power generating
capacity, and 89 additional plants were being built or
planned. Chapter 4 presents a brief picture of nuclear
fuel reserves, describes the slow-neutron reactor types
presently in use for power generation, including their
performance characteristics, and considers the problems
of pollutants and hazards associated with their operation.
Breeder reactor types are then described, and their expected performance is outlined. Finally, in connection
with a discussion of choice among breeders, some of the
main problems still unsolved are presented. The chapter
in its entirety is a short summary of a large problem,
and further condensation here seems pointless. Instead,
a few of the problems most in the public eye will be con-
sidered--the hazards of presently commercial reactors,
the argument for development of breeders, and some of the
hazards and unsolved research problems associated with
breeder development.

The two kinds of pollution of concern are thermal and
radioactive. Thermal water pollution from light-water-
moderated reactors--the kind used predominantly in the
United States and in most other countries--is about 60%
greater, per megawatt of power, than from fossil-fuel
plants. This means that most nuclear plants of the fu-

ture will be located on the ocean or that they will be
provided with cooling ponds or with wet or dry cooling
towers, at an added cost of about 0.08,0.1-0.2, or 0.9-1.2
mill/kWh.

Escape of radionuclides from power plants and fuel-re-
processing plants is held to insignificant levels by multiple barriers, except for the long-lived radionuclides
Kr^{85} and H^3, which escape from reprocessing plants at
rates presenting no present problem. The major hazard of
nuclear power-plant operation is associated with transportation of radioactive cargoes, particularly wastes going to processing plants in sealed stainless steel casks.
The number of casks of spent fuel is expected to rise
from 30 in 1970 to about 9500 in the year 2000. By use
of the Department of Transportation's overall rail accident rate of 0.3 serious accidents per million miles, an
estimated 1.4 serious rail accidents per year is projected
for the year 2000 unless radioactive cargo shipments receive special rail handling. Proposed procedures for disposal of wastes in a salt bed in Kansas have received in-
tensive study and the approval of a National Academy of
Sciences-National Academy of Engineering committee, but
the plan has not been put into action. No better alternative has been suggested.

The projected growth in electrical energy consumption
is so rapid that the U.S. generating capacity, now
300,000 megawatts, is expected to reach 1,500,000 MW by
2000; and nearly half that will be nuclear. The cumulative consumption of uranium ore concentrates by that time
(at the rate of 171 tons per 1000 MW-years) is estimated
at 1.6 million tons if breeders have made no significant
contribution by then. The quantity 1.6 million tons is
to be compared with the estimate, given by the AEC, of
U.S. reserves as a function of uranium price:

Price of U concentrates, $/lb U_3O_8	8	10	15	30	50	100
10^6 tons U resources at this or lower price	0.594	0.94	1.45	2.24	10.0	25.0

According to the table the reserves are there--at a price; and according to the table the price of U_3O_8 will have doubled by 2000 A.D. This is the argument for developing the breeder;* the price of uranium would then be almost immaterial. The ultimate need for breeders is unarguable. With respect to the urgency for their development, however, there is room for much disagreement. Doubling the nuclear fuel cost in the prebreeder era would permit the price of uranium ore to go up more than fourfold, since the fuel cost at the power plant is presently due one-third to ore and two-thirds to concentrating and cladding. The history of fuel-reserve projections is rich with underestimations, and uranium projections may not be exceptional. Uranium prospecting was vigorous a decade ago, but light-water reactors have not become available as fast as was anticipated and the pressure to find new ore is reduced. The state and completeness of our prospecting is perhaps illustrated by the fact that a uranium discovery in Australia last year may be larger than the known U.S. supply. Such events underline the "iffy" character of the AEC projection of price versus tonnage. Prospecting for certain types of uranium deposits such as pitchblende in igneous rock faults has not received the attention it deserves. The mining industry would be enormously stimulated to search and find if the price of crude U_3O_8 were, for example, increased fourfold in association with extended use of light-water reactors. Until

----
*Breeders multiply the energy obtainable from one unit of ore by about 130.

there is a clear cost advantage of breeders over light-
water reactors, independent of projected increases in uranium cost, the extent of the need for hurried development
is no better established than is the knowledge of uranium
reserves.

Three breeder concepts are being pursued. The Liquid
Metal Fast Breeder Reactor (LMFBR) uses liquid sodium to
move thermal energy from the reactor to the point of
steam generation; the Gas-Cooled Fast Reactor (GCFR) is
cooled with helium at 1250 psia. Both are fast-neutron
breeders, doubling their fissionable material in 8-9
years (maybe, later, 5 years for the GCFR). The Molten
Salt Breeder Reactor (MSBR) is a slow-neutron breeder
using a fused salt mixture of uranium and thorium fluorides dissolved in bismuth and lithium fluorides for both
fuel and coolant, and withdrawing fuel continuously for
processing; its doubling time may be as high as 20 years.
Comparison is difficult. The LMFBR has presently the
highest projected performance, but the GCFR may pass it
if a carbide fuel can be developed for higher-temperature
operation. The MSBR has the safety associated with slow
versus fast neutrons, and it uses relatively plentiful
thorium. All three concepts are characterized by major
unsolved problems. For the present favorite, LMFBR,
these include metallurgical problems of cladding to withstand higher neutron bombardment, the hazards associated
with large sodium voids, and the development of on-site
reprocessing if the dangers of shipping very hot waste
high in hazardous plutonium are to be avoided.

If there is a question about the urgency of breeder
development to commercial scale, there can also be a question about the timing of a choice among breeder types.
Either a premature or an overly-prolonged decision can be
costly both in time and in money. Full-scale development
involves a very large annual expenditure. After a deci-

sion has been made on which breeder to develop, a few
years of costly developmental effort makes a change in
path traveled nearly impossible. Hurdles that have
proved to be nearly insurmountable will produce an increase in expenditure rather than a switch to another
breeder concept. To wait, however, until all good nuclear physicists agree on which breeder to back would be
absurd. The AEC provides modest funding for continued
developmental activities on both the GCFR and the MSBR
systems ($5 million/yr each) and directs most of the
available reactor development budget toward the LMFBR program (over $100 million/yr). This emphasis is the direct
result of AEC evaluations of these breeder concepts conducted up to 1969. The LMFBR has also been chosen for
development by U.S.S.R., U.K., and France.

The central issue in the reactor development program
relative to introducing any concept is the ability to
take a feasible system and gather sufficient industrial
and utility support to project it to the demonstration
and commercial plant stage. Such an effort requires
large financial and technical resources. The AEC view-
point is that these resource requirements are such that
the United States can afford to fund effectively only one
concept, the LMFBR, by initial construction of two demonstration plants (with starts spaced about two years apart)
to ensure that there will be more than one commercial
manufacturer of LMFBR plants. However, funds are being
made available to the AEC for only one plant, which supports the AEC contention that the funds available are not
sufficient to permit emphasis on more than one concept at
this time. Supporting the other side of the argument--
U.S. vulnerability in the event of technical-design and
safety difficulties with the LMFBR concept--is the U.S.
record in the reactor development area, which is replete
with examples of failures to succeed in the development

of promising reactor concepts. The ultimate question is
how the prospects for success of the LMFBR--including the
short-term freedom from detrimental technical problems
and the ultimate potential offered by its construction--
compare with the projected ultimate potential of the GCFR
and MSBR.

1.3.4. Central Power from Fossil Fuel

Improved Gas Turbine Cycles. The generation of power
from steam has had centuries of development and has
reached what appears to be a plateau at a thermodynamic
efficiency of about 40%. Higher efficiency has been
achieved, but there appears to be general agreement that
steam temperatures much above 1000 F are not economically
justifiable. The gas turbine-compressor combination is a
World War II baby, developed for aviation use to produce
jet thrust and applied later to various industrial uses,
including peak-shaving in power plants. Stimulated by
the ever-increasing demands for aviation propulsion power,
the aviation industry is continuously raising the inlet
temperature to the gas turbine and the compression ratio
of the air compressor; and these improvements become
available for use in stationary power plants. A recent
study indicated that a gas-steam combination power cycle
holds great promise for efficient generation of power
from coal while simultaneously solving the sulfur pollution problem.

The proposal is to generate producer gas under a pressure of about 20 atmospheres by counterflow of coal and
of air plus steam (cocurrent flow, with coal in pulverized form in second-generation units), scrub the H_2S out
of the gas, burn it at 20 atmospheres, and send it to an
advanced-design gas turbine to make power. The expanded
gases leaving the turbine at atmospheric pressure are hot

enough (1000-1200 F) to generate steam for power in a
waste-heat boiler. The process study has been divided
into first-, second-, and third-generation concepts, corresponding to the early seventies, eighties, and nineties.
It is estimated that power costs from this system would
be 2.67 mills higher, 0.36 mill lower, and 0.31 mill lower than conventional steam plants before addition of
stack-gas cleanup to the latter. (The seeming loss of
ground in the third generation reflects the more stringent control on sulfur by 1990). What is most impressive
is that careful estimates, including allowance for thermodynamic losses, indicate that third-generation gas-steam
plants may be able to achieve an efficiency from gas to
power of 58% with a compressor-compression ratio of 36, a
turbine inlet gas temperature of 3100 F and transpiration
cooling of the large blades that would be used in a 350
MW gas turbine. Combined with the efficiency of 87% for
a third-generation coal gasifier, an overall thermodynamic efficiency of power production from coal is projected
to be 0.58 x 0.87 or 50.4% before a small subtraction for
auxiliaries. Elimination of the waste-heat boiler and
steam turbine gives a base load operation that is less
efficient but much reduced in capital cost and attractive
for use with natural gas or oil.

Magnetohydrodynamics. MHD is a possible technique for
central-station power generation at thermal efficiencies
of 50% or higher. A number of different types of MHD cycles have been considered, but the most likely near-term
hope is an open-cycle fossil-fueled system using an MHD
topping unit. However, MHD requires high temperatures
(4000-5000 F), the use of seed materials to promote ionization, and other troublesome conditions. The fundamental problems in MHD development--gas conductivity, seed
recovery, and materials--remain unsolved despite more
than a decade of effort by numerous competent research

teams in several countries. Now added to these hurdles
are problems introduced by environmental and fuel constraints, including nitric oxide formation in high concentrations and various effects of coal slag. These obstacles leave little room for enthusiasm about the prospects
for MHD development, especially since the contribution
that it might eventually make--electricity from fossil
fuel at a thermal efficiency exceeding 50%--can be
matched by gas turbine developments--in combination with
a steam bottoming cycle--which invoke only a foreseeable
development in blade cooling and anticipated progress in
materials. The appropriate path to improving the prospects for MHD power generation is investment in fundamental, small-scale research in the problem areas mentioned.
Specific research needs are outlined in Section 5.3.

Cryogenic Alternators. The merits of superconducting
electric generators for power-station use appear to be so
outstanding that a brief status report is included on
this subject in Section 5.4.

Central Station Power from Fuel Cells. The concept of
supplying large-scale power by fuel cells has received
attention because, in theory, fuel cells have potential
for highly efficient fossil-fuel utilization with minimal
effects adverse to the environment. In practice, however,
fuel cells do not give the cost and performance characteristics demanded in commercial application. There is no
indication that the outstanding problems involved in
using fuel cells for central-station power production are
close to solution or that significant advancement would
result from large spending on research and development
now. That should await more progress through small-scale,
fundamental research.

1.3.5. Utilization-Related Energy Problems

Automotive Power Plants. The automobile is the major pol-

luter of our air, and an enormous amount of research is
going into improving the gasoline engine or replacing it
by steam or electric power. Chapter 6 presents the present and early-future standards set by the Clean Air Act,
a summary of NAPCA's opinion and evaluation of proposed
technical approaches to emission control, APCO's program,
discussion of replacement of the Otto cycle by gas turbines, steam engines, and electric drives, and comments
on the Wankel engine and the stratified-charge engine.
Space Heating and Cooling. About 22% of the nation's
energy consumption is for space heating. It is probable
that more improvement in heating and air conditioning
will come from use of what we now know about components
and processes than from discovery of new ones; and the
improvement will come as a result of systems studies. It
is in this area that research in heating and air conditioning is in greatest need of being strengthened. A number of examples of need for systems studies in the domestic heating area are given in Section 6.2.4.

1.3.6. Solar Energy Utilization

The emphasis on clean energy in recent years has caused
much comment on the attractiveness of relying on the sun
for power and heat.

Solar energy can be described almost completely by two
numbers, measuring quality and quantity. The quality of
sunlight is almost identical to radiation from a black or
perfect thermal radiator at 6000 K, which is a way of
saying that its thermodynamic potential or theoretical
maximum fractional convertibility into work is extremely
high. Said in still another way, a high fraction of the
energy from the sun is in the form of shortwave radiation,
capable of photosynthesis, of interaction with atoms and
electrons in crystal lattices (reference to photovoltaic
cells), or of coming to equilibrium with high-temperature

receivers through suitable wavelength-selective filters.
The other number of the pair that characterizes sunlight
is the "solar constant," the energy rate onto a unit surface perpendicular to the sun's rays external to the
earth's atmosphere--430 Btu/ft^{2}hr before atmospheric absorption or scatter. This implies that solar energy is
extremely dilute; its flux density onto the earth is only
one five-hundredth of that onto the surfaces of a modern
steam boiler. Unlike most industrial process equipment,
devices to intercept or collect solar energy benefit little by scale increase. Consequently, if solar energy is
to find extensive use, it will tend to be in small units
to accomplish individually small tasks. Domestic hot
water from the sun is economically significant in many
areas today; solar house heating is significant in some,
and its prospects are improving; solar distillation to
produce fresh water from saline water is economic in
areas of extremely high fossil fuel cost; solar electric
power from photovoltaic cells is significant in space research where the laws of terrestrial economics are inap-
plicable, and it has some chance of becoming much cheaper.

The flat-plate collector has been studied extensively
as a device for possible use in house heating or power
production. For the latter purpose an optimized design
using conventional materials yields an efficiency of conversion of solar power to useful power of only 4% in very
sunny areas. There is a possibility of markedly improving the performance by replacing the blackened absorb-
er surface by a "selective" black--one that absorbs the
short waves of the sun but is a poor emitter of the longer waves that characterize radiative loss from the collector. Calculations on such a collector indicate a rise
from 4% to about 10%, still far from interesting in competition with conventional power sources.

Though the prospects of using solar energy as a source

of power are very poor, it has good prospects as a source
of low-level heat--for space heating, food drying, hot
water supply, etc. Enough experimental work has been
done on space heating to furnish a basis for reliable estimations of performance. Optimized designs of solar
house heating systems for eight climatically different
locations in the United States lead to the conclusion
that, except for one location (Santa Maria, California),
house heating with gas or oil is much cheaper than with
solar energy. However, the growing use of electric space
heating indicates that comfort is often not bought on a
cost basis; and solar heating is cheaper than electricity
in seven of the eight locations (Seattle-Tacoma being the
exception).

Photovoltaic cells have efficiencies of conversion of
solar energy to electrical energy of about 12% for the
silicon cell and 4% to 5% for the much cheaper cadmium
sulfide cell. Space power plants based on the silicon
cell have cost $200 to $300 per watt for the cells alone.
The possibility of a cost reduction to $9/watt has been
presented. Polycrystalline cadmium sulfide cells have
better prospects of cost reduction, $5/watt being consid-
ered a possibility in a few years, $1 to $1.50 later; 35¢
and even 10¢ have been projected.

Focusing solar collectors to generate steam for solar
power plants were the favorite device of experimenters of
the last century and the early twentieth century. The
wind forces on a solar collector make the cost of an equa-
torial mount to follow the sun prohibitive.

Photochemical use of solar energy is nature's way.
Outdoor performance of the most efficient plants is not
very impressive--under 3% for the most efficient, chlo-
rella. That number has not been approached within three
orders of magnitude by man's attempts at photosynthetic
energy storage in the laboratory.

1.3.7. Postscript

The fact that no comments have been made on whether man's
appetite for energy should be whetted or curbed does not
mean that no views are held on the subject. Learning how
to use energy more effectively is unequivocably a good
thing, and research directed toward improvements in efficiency of use is to be encouraged. Learning how to supply an ever-increasing per capita demand is a good thing
within limits; we need to watch it.

References

Darmstadter, J., and associates, 1971. Energy in the
World Economy, to be published by the Johns Hopkins Press,
late 1971.

Ritchings, F. A., 1971. "Trends in Energy Needs," talk
at the M.I.T. Alumni Seminar on Providing Energy for the
Future, Cambridge, Mass., May, 1971.

SCEP, 1970. Man's Impact on the Global Environment--As-
sessment and Recommendations for Action, Report of the
Study Group of Critical Environmental Problems, M.I.T.
Press, Cambridge, Mass.

United Nations--World Energy Supplies, 1970. Statistical
Papers, Series J., No. 13, N.Y.

Chapter 2

BACKGROUND FOR ASSESSMENT OF NEW ENERGY TECHNOLOGY

A considerable body of technical information is common to the assessment of the prospects for success of one among several solutions of a fuel-to-fuel or fuel-to-energy conversion problem. Competition among different energy sources for a given end use will depend on relative costs of transporting different fuels, or on fuel transportation versus electric-power transmission. Consequently, this background chapter will include a section on energy transportation costs. Control of SO_2 pollution of the atmosphere, by removal after combustion or by elimination of sulfur from the fuel, is common to various kinds of fuel use; there is consequently a section on sulfur dioxide pollution control. The ubiquitous presence of potential thermal pollution in all fuel-consuming processes makes a section on thermal pollution control necessary. Because pollution by particulates is controllable by fairly well-established technologies and because NO_x pollution control is in a very preliminary state of development, but mostly because time was not available, there is no section on control of these two pollutants. Finally, since the ability to store energy in different ways affects the trade-off among various competing sources of that energy, there is a section on energy storage.

This chapter differs from those that follow it in being concerned almost exclusively with the state of the art; comment on need for change or for new developments is minimal.

2.1. Energy Transportation

2.1.1. Introduction

There is no "best" means of transporting a given form of energy over a given distance except in the context of a specific situation. Oil shipment by tanker may be cheaper per mile than by pipeline, but the water route may be circuitous and result in higher actual costs. Or a coal pipeline may seem to offer an advantage over rail transport, but the rail line may be in place and the mere threat of building a coal pipeline may be adequate to force lower rail rates. Consequently, any general study of energy transportation costs can offer only a general guide to possibilities which are likely to be competitive in a given situation. Emphasis has been placed here on those means of transportation which handle extremely large portions of our energy needs over long distances (railroads, oil tankers) and areas in which technological developments seem most likely to have a significant impact on the pattern of transportation (coal pipelines, distribution of electrical energy); local distribution is mentioned briefly in connection with cryogenic electric power transmission.

In each area, published figures on transportation have been assembled on the common basis of cents per million Btu per 100 miles. This is followed by an identification, where feasible, of major technical or economic factors which affect the transportation costs (e.g., high capital costs, terrain, and capacity in the case of gas pipelines), likely areas for technical progress, and the possible impact of this progress on transportation costs.

The survey of published costs of energy transportation of all kinds is summarized in Figure 2-1 and its accompanying table of explanatory footnotes.

2.1.2. Electric Power Transmission

Examination of Figure 2-1 (top group of curves) leads to such conclusions as: (1) at a given voltage, higher capacity lines reduce unit cost; (2) at a given load level, it is sometimes less costly to use a lower voltage; (3) the unit cost advantage of very high voltages is possible

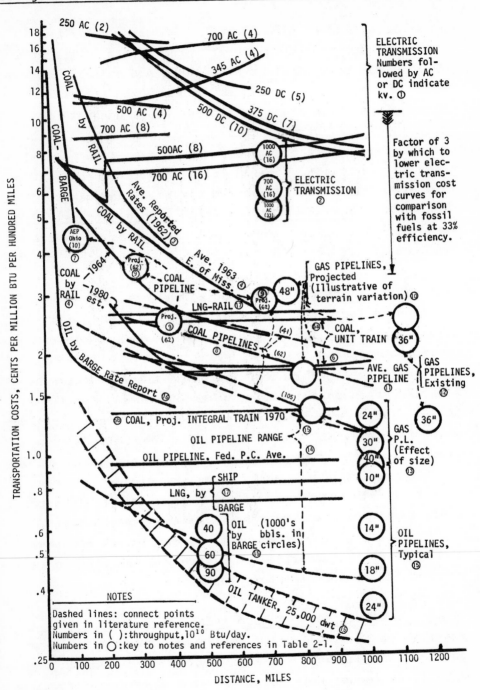

Figure 2-1. Energy Transportation Costs

Table 2-1. Notes and References for Figure 2-1

(References are to small number beside points or lines)

Electric Transmission

1   Point-to-point transmission costs, 85% load factor
    (Federal Power Commission, 1964, pp. 193-203)

2   Point-to-point transmission costs for higher voltages,
    90% load factor (Dillard, 1965)

Coal Transportation

3   Average reported rates by rail for 1962 (Committee on
    Interior and Insular Affairs, 1962, p. 46)

4   Average 1963 rail rates, 1964 volume rates, and 1964
    estimate of 1980 volume rates (Federal Power Commis-
    sion, 1964, p. 60)

5a Average 1970 unit train (Federal Power Commission,
    1970, p. III-3-77)

5b Projected 1970 integral train (Federal Power Commis-
    sion, 1970, p. III-3-77)

6   1964 estimates for unit train (Energy R&D and Na-
    tional Progress, 1964, p. 129)

7   Actual costs, AEP Ohio pipeline (Energy R&D and Na-
    tional Progress, 1964, p. 131)

8   Projected costs as a function of distance for coal
    pipelines of various sizes (Energy R&D and National
    Progress, 1964, p. 131)

9   Projected costs for pipelines from (left to right)
    S. Illinois to Chicago, West Virginia to New York
    City, and Utah to Los Angeles, showing effect of dis-
    tance and terrain on costs (Department of the Interi-
    or, 1962, Tables II-4 and II-7).

Gas Transmission

10 Projected pipeline costs, showing effect of terrain.
    From top: 48" pipeline from Prudhoe Bay to Valdez;
    Ft. Nelson, Canada, to Portland, Oregon; Portland to
    Los Angeles; (at right) Prudhoe Bay to Ft. Nelson.
    All except first case assume 100% load factor (Foster,
    1970)

Table 2-1. Notes and References for Figure 2-1 (cont'd)

Gas Transmission (cont'd)

11 Average cost by pipeline (Federal Power Commission, 1970, p. III-3-77)

12 Existing pipelines in Canada and Canada to U.S. West Coast, showing range of current costs (Foster, 1970)

13 1954 estimates of costs for various pipeline diameters; absolute values are obsolete because of inflation, but relative values are of interest (Gas Engineers Handbook, 1965, p. 8-98).

Oil Transportation

14 Average oil pipeline cost (Federal Power Commission, 1970, p. III-3-77)

15 Ranges of cost for oil transportation in various quantities by various methods (Oil and Gas Journal, 1966)

16 Rates by barge in 1962 (Committee on Interior and Insular Affairs, 1962, p. 146)

LNG Transportation

17 Average costs by barge (Federal Power Commission, 1970, p. III-3-77)

---

only when very large amounts of energy are being transported. Load factor is of course important (most of the curves are based on 85% load factor).

A possibly puzzling point is the flat, rising, or step curves for AC transmission; this is caused by the need to add compensation and control devices as the transmission lines become longer or gain capacity. Such devices are not required for DC transmission.

The two major technical developments which may have a significant impact on long-distance transmission are continued increases in AC voltage levels and introduction of high-voltage DC. Most observers agree that the technical problems involved in building ultrahigh voltage transmission lines in the 1000-1500 kV range are surmountable,

but there is some question about when such systems will
become economical. Technical problems now being attacked
are tower size and configuration, switching surge-level
reduction, insulation (flashovers due to contamination of
insulation are more common than those due to switching
surges), and conductor bundling, with associated corona
and radio noise and response to poor weather conditions.
Although the technical problems associated with UHV are
said to be solvable, at least one observer (Dillard, 1965)
is less sanguine about its economic future. Dillard
(Figure 2-2) estimates that the amount of electricity
that must be transmitted to make UHV reduce costs is so
great that there may be no economic justification for
transmission lines in the 1000 kV range for a long time.

The other major emerging concept in long-distance
transmission is high-voltage DC, already applied to a limited extent abroad. Lack of DC breakers and dynamic controls to regulate power flow and the high cost of converter stations have so far more than outweighed the many advantages of high-voltage DC. Recent developments in
solid-state converters and inverters could change this
situation.

Local Distribution of Electric Power. The two principal
means of distribution of electrical energy locally to consumers now in use or envisioned are various types of underground cable operated at ambient temperatures and special cables operated at cryogenic temperatures.

Underground transmission of electricity, long a common
practice for relatively low-power distribution in metro-
politan areas, has received increasing attention as a
means of transmitting larger quantities of power over
greater distances, largely as a result of increased concern about the impact of overhead lines on environmental
quality and increased need for electrical transmission capability in urban areas. Underground cables face serious

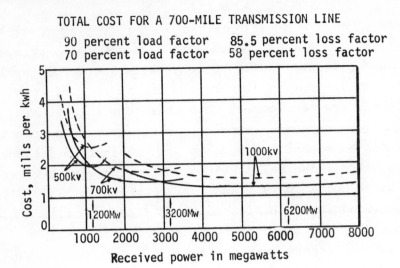

TOTAL COST FOR A 700-MILE TRANSMISSION LINE

90 percent load factor   85.5 percent loss factor
70 percent load factor   58 percent loss factor

Figure 2-2. Total Costs for a 700-Mile Transmission
Line for 500, 700, and 1000 kV, and 90- and 70- percent
Load Factors (Dillard, 1965)

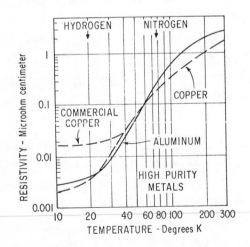

Figure 2-3. Variations of Resistivity with Temperature
at Cryogenic Temperatures (Minnich and Fox, 1969)

problems that are not important for overhead wires: careful insulation of the conductor from the earth, heat removal from the cable, and prevention of corrosion.

Early underground cable consisted of conductor formed
over a hollow core covered on the outside with oil impreg-
nated paper and a lead sheath and filled with oil under
light pressure to prevent voids in the insulation structure. More recently, this type of cable has been super-
seded by the so-called pipe cable, insulated with paper
and tapes and installed three at a time in a section of
pipe filled with oil or nitrogen under high (13-15 atm)
pressure and having no central oil channel. This type of
cable has replaced low-pressure cable largely because
longer lengths are more easily handled. Continuing research and development is in the directions of improving
insulation and increasing cable capacity by cooling.

A somewhat more advanced concept for underground
cables is the compressed gas insulated (CGI) cable characterized, according to one researcher (West, 1970), by
zero dielectric loss, excellent heat transfer characteristics, greatly increased critical length, power capacities equal to overhead lines, uncomplicated and inexpensive terminations, and insulating properties extendable
into the higher voltage ranges contemplated for future
systems.

Cryogenic Transmission Techniques. Work is underway to
develop underground electric transmission cables cooled
by liquid nitrogen, hydrogen, and helium. Such systems
are designed to take advantage of the much reduced resistivity of conductors which occurs at very low temperatures (Fig. 2-3). The price that must be paid for the

increased conductivity is, of course, the cost of refrigerating the transmission lines to extremely low temperatures. Table 2-2 illustrates the limits that thermodynamics and the practical capabilities of cryogenic refrig-
erators place on the advantages that can be obtained
through cooling. It can be seen that the gain in conductivity with liquid hydrogen relative to the refrigeration
costs is greater than is the case with nitrogen. Other
factors are the economics of vacuum-installed piping and
of refrigeration, dielectric losses and, under AC conditions, significant additional losses not included in the
table.

Economic estimates are difficult and questionable
since no commercial line of any sort is yet in operation.
Because 90% of the cost of present underground cables is
for capital and only 10% is for losses, it is possible
that when and if cryogenic transmission systems become
economic they will be characterized by higher power

Table 2-2. Cryogenic System Parameters (Minnich and Fox,
            1969)

Liquid	Operating Temperature (Degrees Kelvin)	Metals Relative Resistivity	Refrigeration Ratio* Watts Input/Watts Load	
			Theoretical	Practical
Oil	293	1	1	1
Nitrogen	77	1/8	3	6-10
Hydrogen	20	1/500	14	40-100
Helium	4	Supercon-ductors	75	300-1000

*Last two columns = (refrigeration load)/(I^2R transmis-
sion loss), proportional to refrigeration if I^2R is maintained constant among coolants.

losses than present systems, with capital costs per MVa-
mile kept low by much higher power capacities than those
of present underground transmission systems.

There is disagreement as to which type of cryogenic
cable is most likely to see early service. Graneau (1970)
offers comparisons of the costs involved in using different coolants and conventional overhead and underground
lines based upon various assumptions concerning load
(Figure 2-4).

Another view is that hydrogen may offer a better com-
promise between the advantages of low temperature and the
costs of cooling. Corry (Papamarcos, 1970) envisions the
following lineup of technologies for minimum-cost underground transmission in the future: 300-500 MVa: conventional systems with taped cables and perhaps extruded di-
electrics; 400-1000 and perhaps up to 2000 MVa: CGI cables; 1000-3000 MVa: resistive cryogenic cable; and 3000

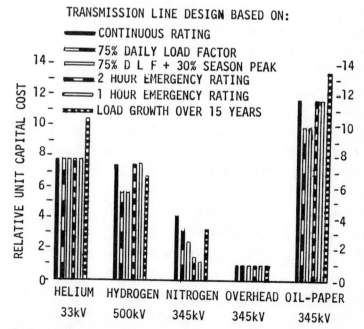

Figure 2-4. Underground Relative to Overhead Transmission
Cost (Graneau, 1970)

MVa up: superconducting cryogenic cable. However, he
foresees no breakthrough which will make underground
transmission systems economically competitive with familiar overhead systems.

2.1.3. Gas Pipelines

Figure 2-1 (circles right-center) indicates that the cost
of long-distance transmission of gas varies from 1¢ to 3¢
per million Btu per 100 miles. Costs are relatively in-
sensitive to distance beyond the 80 or 100 miles that
compose a line section, but economies of scale are very
significant. The effect of pipe diameter is illustrated by three 1000-mile points for 24-, 30-, and 40-inch
pipe.

Pipeline installation costs are sensitive to the remoteness and roughness of the terrain, as indicated by
the estimate of 3¢$/(10^6$ Btu$)(100$ mi$)$ for the 48" trans-
Alaska pipeline and the 1.35¢ estimate for a line from
Portland, Oregon, to Los Angeles.

Table 2-3 gives a representative cost breakdown for
natural gas transmission. Since 70% to 80% of the cost
represents pipeline fixed costs, with the remainder divided between the fixed and operating costs of the compressor stations, it is clear that maintenance of a high load
factor is important to reduction in unit cost; distribu-
tors will sell off-season interruptible gas at bargain
rates.

One other variable that has a modest effect on transmission costs is the heating value of the gas. Hunsaker
(1966) found that the compressor horsepower required for
a constant volumetric throughput increased as heating
value increased, but that the horsepower required per
unit of energy delivered decreased. However, over the
range of heating values from 1032 Btu/cu ft to 1248 Btu/
cu ft, this effect was relatively small, with required

Table 2-3. Representative Cost Breakdown for Natural
Gas Transportation (Energy R&D and National
Progress, 1964)

Costs	Percent of transportation costs	Costs	Percent of transportation costs
Fixed costs:		Operating and maintenance costs:	
Depreciation (3.3% of depreciable property).......	17.8	Pipeline operations...........	5.6
Ad valorem taxes	7.0	Administration..	5.7
Interest 5-1/2% (on 65% of investment)...........	19.7	Fuel gas and losses..........	10.7
Balance to capital stock...	16.1	Total	100.0
Federal income taxes, 52%......	17.4		

Btu rising only 4% for constant volumetric throughput and
falling 12% for constant energy throughput.

Efforts presently underway to develop economical synthetic gases are concentrating on finding a process for
producing a gas which is interchangeable with natural gas,
primarily because consumer equipment has been designed to
use natural gas and would have to be converted to use a
lower Btu substitute. Present pipeline technology would
be suitable for transportating a lower-Btu substitute
gas, however.

Table 2-3 indicates that the chief prospect for reduction in transportation cost lies in capital savings associated with development of lighter and stronger pipe and
improved construction techniques. Lower-cost compressors,
however, could make a contribution. A significant trend
is the movement toward gas-turbine-driven centrifugal
compressors from the traditional engine-driven reciprocating units. In 1970 for the second consecutive year,

more gas-turbine horsepower was installed than reciprocating. A major factor in this change is that gas-turbine
horsepower costs $264/hp for new and $210 for additions,
versus $492/hp for new reciprocating units and $325/hp
for additions.

2.1.4. Transportation of Liquefied Natural Gas (LNG)

In the United States, LNG is used primarily for peak sha-
ving and in situations in which normal gas transmission
and distribution is either unavailable or uneconomical.

The technology for shipping LNG by truck is well in
hand, having been developed for commercial liquefied
gases and for handling liquid gases in the aerospace program. Much more economical for long-distance shipment
are cryogenic railroad cars. However, these are not competitive with natural gas pipelines for shipment of large
quantities of LNG over long distances.

Use of pipelines for LNG has been considered and re-
jected for long-distance shipments; insulation and refrigeration presently cost too much. There still must be developed an economical inner pipe and insulation system
which can be transported from the fabricator to the right-
of-way and installed easily. The question of number of
pumping and refrigeration stations needed on long-distance LNG pipelines is undecided; one published figure is
every 25 miles, while another report recommends every 100
to 150 miles. There are attractive gains to be made if
an LNG line is developed; LNG requires but a tenth the
pumping horsepower of equivalent gas. High-volume LNG
movements over short distances have generally been
through low-temperature pipelines.

The design of LNG tankers has been developing rapidly
since the first prototype went to sea in 1959. Earlier
designs used self-supporting insulated tanks which were
inserted in the hull after fabrication in a shop. More

recently an attempt has been made to improve tanker hull
space utilization and reduce costs through introduction
of the "membrane" concept. In a membrane tanker, self-
supporting tanks for the LNG are replaced by a thin "membrane" of cryogenic material which is supported by the
insulation and hull and is molded to closely follow hull
contours.

LNG tankers for the most part use conventional steam-
turbine propulsion systems modified to use the gas boil-
off from their tanks. However, because of the low density of the LNG cargo, they must have special ballast systems. Tanker size has been increasing steadily as more
experience is gained in handling this unconventional cargo. While LNG tankers do not approach recently-construc-
ted oil tankers in size, continued size increases in LNG
tankers are expected. Figure 2-5 shows that the savings
from increased tanker size increase less rapidly than
tanker size.

LNG travels at atmospheric pressure and -259 F. Because the cost of refrigeration increases rapidly with
decreasing temperature, other marine gas transport concepts have been developed. Two of these are medium-con-

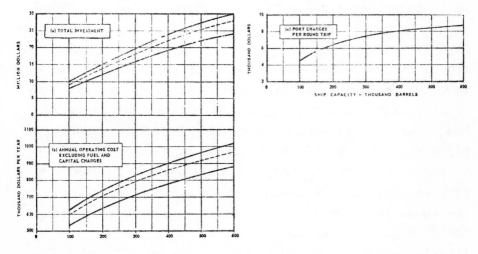

Figure 2-5. LNG Carrier Costs (Institute of Gas Technology, 1969)

diton liquefied gas (MLG) and compressed natural gas
(CNG): MLG requires 200 psig and -180 F, while CNG
needs 1150 psig and -80 F. MLG requires about half the
refrigeration horsepower needed by LNG, while CNG needs
only a quarter. Both require containers that can withstand the higher pressures than used with LNG, though
these materials need not have equivalent low-temperature
properties.

Barges have not been a significant means of transporting LNG. However, the technology required is similar to
that already in use for trucks and railroad cars, and
barges are in use to haul other cryogenic materials.

2.1.5. Oil Pipelines

Pipelines are generally the least expensive way to move
petroleum and petroleum products overland. According to
the Federal Power Commission the average cost for moving
oil through pipelines is around $1.0¢/10^6$ Btu per 100
miles. In a given situation this may be as high as 1.6¢
or as low as 0.4¢, however. Two factors which play an
important role in pipeline economics are enonomies of
scale and the load factor. The potential for reducing
costs through the use of larger diameter pipe is illustrated by points in Figure 2-1 indicating costs for 10-,
14-, 18- and 24-inch pipelines. Increasing pipe diameter
from 10 to 24 inches can reduce unit cost of transportation by 60%. Like gas pipelines, oil pipelines have extremely high fixed costs, usually between 70% and 80% of
the total. Wages account for only 15%. As a result,
pipelines are operated to ensure a high load factor.

It has been estimated (Committee on Interior and
Insular Affairs, 1962) that 40% of U.S. oil pipeline mile-
age consists of crude oil gathering lines, 34% of crude
oil trunklines to refineries, and 26% refined product
trunklines.

Recent technical progress includes the use of pipeline diameters up to 42", greater use of pipelines in hostile environments such as underwater and in polar regions, and increases in strength which allow higher pressures. Friction and corrosion-fighting coatings are also being developed. Some areas of greater challenge in which there is potential for technical progress include detection of interfaces between product batches in pipelines to permit full automation of the line, development of successful insulation techniques to permit pipelining of heated heavy fuel oil, and the development of automatic welding techniques to reduce this important component of construction costs. In addition, the use of computers to control pipelines is being advanced.

2.1.6. Petroleum Tankers

Costs. Figure 2-1 shows that, nuclear fuel excepted, there is no cheaper way to move energy over long distances than by oil tanker. The fairly wide variation in unit cost is due to a number of factors: Tankers are owned by a variety of oil and nonoil interests, and there exists both a spot- and a long-term charter market; and these factors tend to complicate the cost situation. Tanker rates vary over time in response to world supply and demand and in response to such events as the closing of the Suez Canal or the more recent closing of the Trans-Arabian Pipeline. Private tanker owners strive to take advantage of high charter rates whenever possible, while the oil companies endeavor to keep rates down and utilize their own tanker fleets to gain leverage toward that goal. Labor accounts for about 30% of the total cost and, because of U.S. labor and government policies, increasing numbers of U.S.-controlled tankers are registered in countries with lenient regulations--such as Panama,

Liberia, and Honduras. Finally, it should be noted that the time spent in port is an important part of total transportation costs and should be held to a minimum. This consideration, along with others such as the fact that smaller tankers tend to be used for shorter runs, accounts for the downward trend in transportation costs as the length of the run increases.

The data presented in Figure 2-1 are for 25,000 dwt tankers typical of what might be expected on the U.S. coastal trade. High and low ranges are indicated. The supertankers, an order of magnitude bigger than this size, are becoming common in international trade. Cost of transporting oil in supertankers over distances greater than can be represented on the graph are shown in Table 2-4. These figures represent middle- to long-term charter rates.

Technology. Clearly the most spectacular technological improvement in tankers has been their tremendous increases in size, from the 16,800 dwt T-2 standard tanker of World War II to recent designs of several hundred thousand tons. Some indications of the great potential for further economies of scale can be gained from Table 2-1 and Figure 2-1. Recent figures (Cabinet Task Force on Oil Import Control, 1970) indicate that reductions on

Table 2-4. Costs of Long-Distance Transportation of Oil by Tanker (Cabinet Task Force on Oil Import Control, 1970)

| Trip | Cost, $\mathcal{c}/(10^6 \text{Btu})(100 \text{ mi})$ | |
	Present	1980 est.
Persian Gulf-New York or Gulf (11,800 miles)	0.104	0.064
Venezuela-New York or Gulf (1,800 miles)	0.212	0.148
Gulf Coast-New York (2,500 miles)	0.300	0.180

the order of 30% over current figures may be expected by
1980 on medium-length voyages such as from Venezuela to
the United States, and perhaps 40% on longer hauls such
as between the Persian Gulf and the United States.

However, larger tankers have brought with them a series of problems and related technical advances: (1)
They cannot use many harbors because of their tremendous
draft. (2) They must bypass the Suez and Panama Canals
for long-distance hauls, but this is economical and has
not slowed the growth in size. (3) The loading and unloading problem has forced the development of offshore
loading facilities, designed to minimize time lost in
port. (4) The large new ships have been automated to the
greatest extent possible in order to cut the rather substantial cost of the crew.

There is some indication, however, that a point of
diminishing returns has been reached, at least temporari-
ly, in the process of reducing the cost per dwt in huge
tankers by continued increases in tanker size. The following data suggest that tanker size may settle down, for
a while, in the 250,000 ton size (Chemical and Engineering News, 1969):

Size of ship (thousand tons)	35	100	207	250	315	372
Cost per ton (constant $)	135	91	78	65	67	66

This size is further favored by its not being too large
to enter a number of important port facilities.

There seems to be little question that larger tankers
are technically feasible. There are presently yards in
Japan capable of assembling 500,000 ton tankers using the
"assembly line" technique in which prefabricated sections
of 100 to 200 tons are assembled in shops and then welded
into place along the keel.

Other new concepts which have been proposed for oil

shipment over water include the icebreaker tanker and the
submarine tanker, both pertinent to the problem of moving
Alaskan oil to market through Arctic regions. A prototype icebreaker tanker, the Manhattan, has successfully
made the trip from New York to Prudhoe Bay. The development of these techniques will, of course, depend upon the
success or failure of various other schemes to move Alaskan petroleum, such as a pipeline across Canada or pipeline shipment to Valez followed by tanker shipment to the
U.S. West Coast.

The increased size and number of oil tankers plying
the world's waterways poses some new pollution problems
of unprecedented magnitude if a 300,000 dwt tanker were
to sink. The possibilities of such a disaster are increased by the fact that the monster tankers are under-
powered and much less maneuverable than smaller ships, as
well as the fact that their greater draft requires that
they be more careful of underwater hazards.

2.1.7. Coal Transportation by Rail

The gradual rise in freight rates for coal following
World War II came to a halt in the late fifties and early
sixties as increased competition from fuel oil shipped
over water to the east coast and the potential for building coal slurry pipelines to the east coast from coal
fields brought an end to regulatory and technological
stagnation in the railroad industry.

The series of curves (Figure 2-1) representing coal
transportation costs for successive time periods illus-
trate the gradual lowering of railroad rates for larger
quantities of coal following liberalized regulation and
the introduction of new technologies. The most advanced
railroad hauling technology now in regular use is the
unit-train concept, which consists of conventional equipment operating continuously in a train dedicated to the

service of one customer. This concept can increase the
utilization of cars by as much as 600% and reduces the
costs of loading, unloading, and switching associated
with coal shipment as part of a general train carrying
many different types of goods. Often the utility receiving fuel owns the cars which comprise the unit train.
This decision of course locks the utility into a certain
mode of transport for a time, but may reduce overall
costs. Table 2-5 presents a hypothetical breakdown of
costs for a unit train.

One other aspect of transportation costs incurred
through use of unit and integral trains (described below)

Table 2-5. Estimated Unit-Train Freight Rates, in Cents
           per Ton (Energy R&D and National Progress,
           1964)

Cost item	Distance, miles			
	250	500	750	1,000
Out-of-pocket cost:				
Expenses:				
Wages[1]	7.5	15.0	22.5	30.0
Operating expense	52.5	105.0	157.5	210.0
Return on investment:[2]				
Freight cars	1.6	2.8	4.0	5.2
Locomotives	1.2	2.1	3.0	3.9
Road properties	12.8	25.6	38.4	51.2
Total out-of-pocket cost and return on investment	75.6	150.5	225.4	300.3
Constant costs[3]	50.0	82.5	115.0	147.5
Fully distributed costs	125.6	233.0	340.4	447.8

[1]Based on 1961 wage scale.
[2]Provides a 4-percent rate of return on investment after
income tax.
[3]Includes distribution of 50% rate of return on road
property (freight), overall revenue, tons and ton-miles.

is that of improving the coal-handling facilities at the receiving plant to achieve higher rates of unloading.

The next step in reducing costs is the construction of trains specifically designed for shuttle use, dubbed integral trains. Such trains would consist of extra-large cars semipermanently coupled, with power units interspersed among the cars, and specially designed for rapid loading and unloading, with features such as swivel couplers to allow the entire car to be rotated for unloading. With power units at either end, the trains would not need to be turned around at terminals. Various control aspects of unit-train operation would be automated to the greatest extent possible. Of course, extremely large amounts of coal would have to be involved for this train to be economical. It would have a capacity of about 35,000 tons, compared with 10,000 tons or so for current large trains. The possible impact of such a train should be evident from the cost curves for integral trains. Although the reliability of the projected costs is probably not high (the curves should drop with increasing distance), the numbers indicate that the integral train should be competitive with coal pipelines and even gas pipelines. This would indicate that the choice of mine-mouth versus load-center coal gasification plants might be difficult to make and depend on a better analysis of integral train operation if transportation costs are an important component of overall costs at the city gate.

It should be remembered that railroads' flexibility and hence ability to institute new procedures is hindered by rate regulation, labor problems, uncertainty caused by mergers, etc. However, such barriers have been overcome to some extent in the past when action was forced by competitive pressures.

2.1.8. Coal Pipelines
The only commercial coal slurry pipeline which has been

operated in this country (prior to the recent opening of
the line from Black Mesa to the Mohave) was built in Ohio.
It began service in 1958 and continued to run through
1963, when it was closed after railroads in the area reduced their rates for coal. Consequently most of the
cost estimates for large slurry pipelines are only projections, and are not based on actual experience. Three
dashed lines of Figure 2-1 indicate a significant drop
expected in unit cost as the length and throughput increase, due primarily to the considerable cost contributed by such items as slurry preparation (Table 2-6). It
is to be noted that two of the points on Figure 2-1,

Table 2-6. Capital Investment, Coal-Water Slurry Pipeline (Department of the Interior, 1962)

| | Cost, Millions of Dollars | |
	Utah to Los Angeles- San Diego Area	Southern Illinois to Chicago Area
Slurry preparation	$ 15.40	$15.40
Gathering	3.20	2.91
Pipeline Cost:		
Main line	95.96	37.59
Laterals	46.79	8.72
Storage:		
At preparation plant	1.80	1.80
At power plants	2.80	3.50
General plant	2.75	2.60
Pumping	19.83	7.12
Subtotal	188.53	79.64
Working capital	8.47	5.36
Total	197.00	85.00

based on Table 2-1, correspond to considerably higher
unit costs than the dashed lines referred to earlier.

The operation of the Ohio pipeline is generally
thought to have proved coal transportation by pipeline to
be workable and slurry preparation to be feasible, but it
did not prove out any method of slurry utilization which
was deemed satisfactory. Figure 2-6 illustrates the combination of vacuum filtering and thermal drying which was
used.

Attempts have also been made to use slurry directly to
fire a cyclone boiler. These tests proved to be successful, experiencing a reduction in boiler efficiency due to
moisture in the fuel of only 4%.

A third method of slurry utilization which has been
tested and adopted for the Mohave project is the use of
centrifuges to dewater the coal. Dewatering by this
means to 15% water content is necessary before combustion
is feasible.

It has been found that a slurry of 55%-58% moisture-
free solids is the best mixture judged by solution stability during shutdown, low pressure drop per mile, and
high concentration. The maximum size coal to be used is
No. 8 mesh. Since the coal for slurry is ground finely,
ash and pyrites are liberated. Tests have indicated that
boiler corrosion and air pollution from slurry are no
worse than that from dry coal.

It has been suggested that oil rather than water be
used for slurrying, to increase the energy throughput of
coal slurry pipelines. While this might be attractive
if coal and oil were found together and shipped to the
same destination, the costs of moving oil to the coal
fields makes this economically unattractive. In addition,
oil slurry is more difficult to handle.

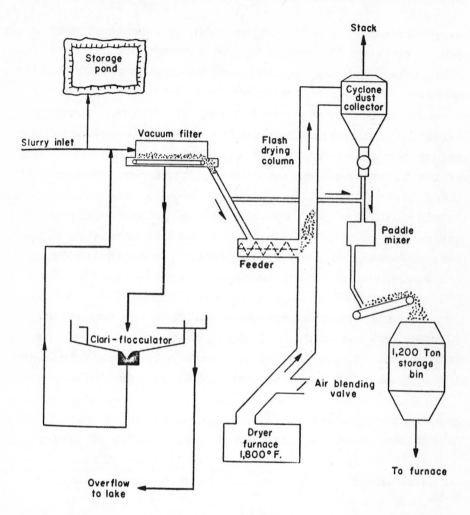

Figure 2-6. Coal Slurry Dewatering Plant, Eastlake, Ohio
(U.S. Department of the Interior, 1962)

2.2.   Sulfur Dioxide Pollution Control

2.2.1.   Nature of the Sulfur Dioxide Problem

The mass ratio of sulfur to carbon released in gaseous
form by the world's fossil fuel combustion is in the

neighborhood of 1 to 100;* the corresponding volumetric ratio of SO_2 to CO_2 added to the atmosphere is about 1 to 250. Organic decay is estimated to contribute about 20** times as much CO_2 to the atmosphere as combustion. The assumption that organic decay adds CO_2 and H_2S (quickly oxidized in the atmosphere to SO_2) in the same molal ratio as the combustion of fossil fuels is a poor one, but of the right order of magnitude. The global concentrations, in the atmosphere, for SO_2 and CO_2 are about 0.3×10^{-9} and 320×10^{-6} by volume, i.e., the two are in a ratio of 1 to 1 million. This is to be contrasted with the feed ratio of 1 to 250. Clearly, the life of SO_2 in the atmosphere is about three orders of magnitude shorter than that of CO_2.

Various mechanisms may be postulated for removal of SO_2 from air to maintain its known low equilibrium concentration. An uncritical and certainly incomplete listing of possible mechanisms would include the following:

Removal by solution in falling rain, snow, and sleet

Absorption on particulate matter and removal of latter by rain

Oxidation to SO_3 by ozone, and solution of SO_3 in rain water

Catalytic oxidation on particulate matter, such as vanadium-containing ash

---

*World carbon consumption in 1968 was about 4×10^9 metric tons. The U.S. energy consumption was 34.8% of the world total. Assume U.S. carbon consumption was in the same ratio. SO_2 discharge into U.S. atmosphere in 1968 = $(22.9 + 7.3) \times 10^6$ metric tons from combustion and industry.
$$\frac{\text{Sulfur}}{\text{Carbon}} = \frac{30.2 \times 10^6 (32/64)}{4 \times 10^9 \times 0.348} = 0.0108$$

**Terrestrial and oceanic production rates of organic carbon are 56 and 35×10^9 (range of variation 22 to 151×10^9) tons/year (SCEP, 1970), a total of 91×10^9 compared to 4.2×10^9 coming from the combustion process.

Reaction with atmospheric ammonia

Reaction with alkaline particulate matter

Several of these could doubtless be dismissed by application of known thermodynamics or kinetics. The correct
mechanism, possibly or probably not listed above, needs
support by experimental research and by integration into
a complete sulfur cycle.

Detrimental effects of SO_2 on vegetation, materials,
and human health are first noticed in areas having an annual mean atmospheric SO_2 concentration of 0.03 to 0.04
ppm (NAE-NRC, 1970). Since this value is 100 times the
average global value and, as noted, above, the rate of
SO_2 emission from biological processes is 20 times that
from fuel combustion, it is clear that combustion-generated SO_2 may present a dispersion problem or a local pollution problem, but it is not a global pollution problem.
In fact, SO_2 concentrations are near and at certain times
they exceed the 0.03-0.04 ppm level in several urban
areas of the United States (NAPCA, 1969a). It is therefore clear that local or regional SO_2 pollution can be
serious in the absence of control measures. Therefore,
national air quality standards have been adopted which
state that the annual arithmetic mean SO_2 concentration
not to be exceeded is 0.03 ppm and that a maximum 24-hour
concentration of 0.14 ppm should not be exceeded more
than once a year (Environmental Science and Technology,
1971). Several localities have set limits on the sulfur
content of power-plant fuel. In some cases, a 1% fuel
sulfur limit has been found to be inadequate, and lower
limits are being considered. The Environmental Protection Agency is expected soon to announce limits on the
SO_2 content of stack gases. The near-term or first-
generation limit is expected to be about 1000 ppm, and

the second- and third-generation limits (for the early
1980s and 1990s) may be as low as 200 ppm and 50 ppm,
respectively.* The limit expected for 1975 is 600 ppm.

The question of how to satisfy the forthcoming SO_2-
emission limits is an important part of the broad national problem of fuels and energy management. Coal and oil
combustion now contributes about 77% of the man-made SO_2
emission in the United States (65% from coal, 12% from
oil), with about 55% of the total coming from power
plants. In view of the rapidly growing demand for electric power, two-thirds of the total SO_2 emissions in 1980
are expected to originate from this source (Table 2-7).

One authoritative projection indicates that the 1970
demand for electric power, about 340×10^6 kW, will double
by 1980 and exceed 10^9 kW by 1990 (Figure 2-7). The use
of oil is projected to increase until its forecasted

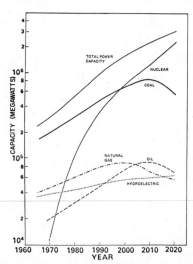

Figure 2-7. Projected Power Generating Capacity and Fuel
Sources of Electric Utilities in the U.S. (with breeder)
(NAE-NRC, 1970)

----

*If the fuel is burned with 10% excess air these limits
are equivalent to sulfur contents of about 450, 90, and
20 $g/10^6$ Btu, corresponding to about 1.2, 0.25, and 0.06%
sulfur in coal.

Table 2-7.   Estimated Potential Sulfur Dioxide Pollution
in the United States without Abatement[a]
(NAE-NRC, 1970)

Source	Annual Emission of Sulfur Dioxide (millions of tons)				
	1967	1970	1980	1990	2000
Power Plant Operation (Coal and Oil)[b]	15.0	20.0	41.1	62.0	94.5
Other Combustion of Coal	5.1	4.8	4.0	3.1	1.6
Combustion of Petroleum Products (Excluding Power Plant Oil)	2.8	3.4	3.9	4.3	5.1
Smelting of Metallic Ores	3.8	4.0	5.3	7.1	9.6
Petroleum Refinery Operation	2.1	2.4	4.0	6.5	10.5
Miscellaneous Sources[c]	2.0	2.0	2.6	3.4	4.5
Total	30.8	36.6	60.9	86.4	125.8

a. February 1970 NAPCA estimates, excluding transportation.
b. See Figure 2-8, with breeder.
c. Includes coke processing, sulfuric acid plants, coal
refuse banks, refuse incineration, and pulp and paper
manufacturing.

availability falls off, around the year 2000. Natural
gas usage is also expected to increase, but to fall off
sooner. Hydroelectric sources are expected to remain
relatively small.

Nuclear generating capacity is increasing rapidly, but
its impact will probably not prevent the power industry
from being primarily dependent upon coal until the end of
the century. Consequently, projections can be made for
sulfur dioxide emissions from power plants, if no abatement procedures were used (Table 2-7 and Figure 2-8).

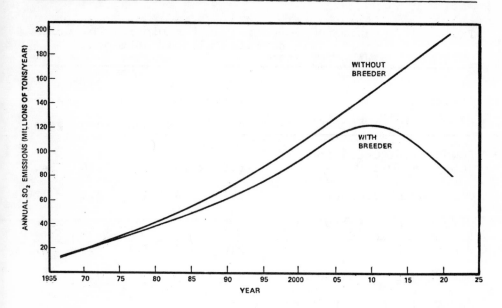

Figure 2-8. Comparison Between Projections for Total Power Plant Uncontrolled Sulfur Dioxide Emissions (NAE-NRC, 1970)

The emission rate in 1990, if uncontrolled, might be over three times the present rate.

The technology for control of SO_2 emissions from combustion processes was recently studied by a National Academy of Engineering - National Research Council panel (NAE-NRC, 1970). The panel concluded that "...contrary to widely held belief, commercially proven technology for control of sulfur dioxides from combustion processes does not exist." This condition, in view of the stated projections, points up the high importance of research and development that bears on controlling SO_2 emissions from combustion processes. Each of the following abatement procedures offers a certain type of potential for the future: (a) the use of fuel naturally low in sulfur; (b) fuel desulfurization; (c) sulfur removal during combustion; (d) stack-gas cleaning; and (e) stack-gas dispersion. It should be obvious that no single approach can be best in all situations, and some cases may require

more than one method. The status of each approach is
discussed below.

2.2.2. Use of Fuel Naturally Low in Sulfur

The substitution of low-sulfur fuels, now the only available method for reducing SO_2 emissions, is only a short-
term solution because of the limited supply of clean fuel.
Because of tightening emission standards, the 1% sulfur
coal that is now adequate for most sulfur limits will
soon be unacceptable. For example, the New Jersey limit
on bituminous coal will drop from 1% to 0.2% sulfur in
October 1971. Even coal less than 1% in sulfur is in
short supply, and the supplies of natural gas and low-
sulfur oil are still smaller.

One-third of the U.S. coal reserves are east of the
Mississippi. Although about 83% of the relatively low-
sulfur reserves are in the West in large deposits of
subbituminous coal and lignite, 95% of the U.S. coal is
mined east of the Mississippi. About 36% is coal of less
than 1% sulfur, mined in the East, and of that about 26%
is exported and 49% goes to metallurgical use. Therefore
only about 9% of the present coal production is low-sulfur coal available for general use.

The cost of low-sulfur coals is at least $2.00/ton
more than that of high-sulfur coals of comparable rank
and ash content. This difference amounts to about 0.8
mill/kWh in power cost.

2.2.3. Fuel Desulfurization

The technology for sulfur removal from distillate oil is
well established, but residual fuel oils are more difficult to treat because they contain metals which deposit
on the solid catalysts employed. The petroleum industry
has invested heavily in the development of ways to desul-
furize fuel oil, and there is little doubt that some of

these methods will work. The cost of reducing the residual fuel oil from 2.6% to less than 1.0% sulfur will
probably increase the price to the power station by 50¢
to 80¢/bbl, which represents a 20% to 35% increase in
fuel cost, or an increase of 0.8 to 1.3 mills/kWh in
power cost.

The three common forms of sulfur in coal are organic
sulfur, pyrites or marcasite, and sulfates (Ca and Fe)
(Ode, 1963). Elemental sulfur in very small amounts has
been found in some coals, but in rare cases as much as
0.15% (by weight of coal) has been reported. Sulfates
rarely exceed a few hundredths per cent except in heavily
weathered or oxidized coals. Marcasite and pyrite, two
crystal forms of FeS_2, are usually referred to as "pyrite." The organic sulfur, present in complex organic
compounds, only rarely exceeds 3% (by weight of coal) but
it may be as high as 11%. The ratio of organic to inorganic sulfur oxides varies widely but is of order unity
for many coals. Usually only about half the pyrite can
be removed by conventional grinding and washing; removal
of organic sulfur can be achieved by hydrogenation in
liquefaction or gasification processes.

NAPCA's survey of naturally occurring low-sulfur coals
and washability tests of coals available for uses other
than metallurgical suggest that of the steam-coal production 8% has 1% or less sulfur as mined and could be
cleaned further, 1% has over 1% sulfur and is easily
cleaned, and 6% has over 1% sulfur and is cleanable at
higher cost (NAE-NRC, 1970). Thus, about 25% of the
steam-coal production could be reduced to a maximum sulfur
content of 1%. NAPCA further estimates that the sulfur
content of the remaining 75% of the steam coals could be
reduced as much as 40% by refinement and broader application of coal-cleaning technology. These figures, together with expected sulfur limits of considerably less than

1%, indicate that coal beneficiation alone offers only a
small potential for controlling SO_2 emissions.

A more promising approach is coal gasification accompanied by sulfur removal. This approach has a number of
advantages including the following: (1) Sulfur removal
is accomplished prior to combustion when the volume of
gases to be treated is considerably less than that of the
stack gases; (2) absorption techniques for removing H_2S
from the product gas are now available; (3) the clean gas
permits use of more efficient power cycles, such as gas
turbine or combined gas turbine and steam turbine cycles
(Section 5.2) and perhaps MHD in the more distant future
(Section 5.3), thereby offsetting the cost of desulfurization.

Several U.S. processes for making a high-Btu natural
gas substitute from coal are at various stages of development, the most advanced being at the pilot-plant scale.
The projected gas prices are in the range 55-68¢/10^6 Btu
(this is probably optimistically low; see Section 3.1).
This development work is an effort to improve on the German Lurgi process, now commercially available, which is
believed to be capable of making high-Btu gas in the
United States at a price of about 90¢/10^6 Btu (Tsaros,
1971).

Clean low-Btu gas can be made from coal using a gas
producer followed by a gas purification system. One such
process, using a Lurgi gasifier, is now supplying fuel to
a gas-turbine steam-turbine power plant in Germany, and
other such processes have recently been studied in the
United States (Section 3.2). The projected low-Btu gas
price for a Lurgi-based system using 20¢/10^6 Btu coal in
the United States is 55¢/10^6 Btu.

The projected increase in power cost attributable to
fuel desulfurization via coal gasification depends upon
the power system considered. Among the many alternative

schemes are (1) a low-Btu gasmaker close-coupled with a
power plant (a) at the mine mouth with electrical transmission, or (b) near the center of use with coal transportation, and (2) high-Btu gas production at the mine
mouth with gas transmission to a power plant near the
center of use. Although a full evaluation of the impact
of desulfurization on power cost should consider all
parts of the power system affected by the gasification-
desulfurization step, such as energy transmission (Section
2.1), a simplified analysis given in Section 3.2 is in-
structive. The results indicate that a switch from coal
to clean low-Btu gas in a conventional power plant (38%
efficient) would increase the power cost by about 2.6 to
2.9 mills/kWh. The cost increase is reduced to 1 mill/
kWh if the gas is burned in an advanced-cycle gas-steam
power plant having 45% efficiency (Section 5.2). The
predictions further indicate that the sulfur limit expected 10 years hence can be met with a second-generation,
pressurized low-Btu gasifier and gas purification system,
and that, if such a system is coupled with an advanced-
cycle gas-steam plant, the cost of power will then be
1 mill less than by use of a conventional steam system
operating on coal. All three of these comparisons need
adjustment to cover the cost of SO_2 removal from the
stack gas of the conventional system. If that cost is
more than 1 mill per kWh, power from the first-generation
advanced-cycle gas-steam plant will cost less than conventional steam power.

The production of low-sulfur liquid fuels from coal
has also been studied (Section 3.3), and this process
looks attractive.

2.2.4. Sulfur Removal during Combustion
Two new combustion processes which offer the possibility
of being brought to commercial acceptability in the later

1970s achieve sulfur dioxide control by burning the coal in the presence of a sulfur acceptor. These processes require changes in boiler design, manufacturing proce-dures, and operation, and their acceptance by boiler manufacturers, utilities and industry may be slow unless they receive sufficient support to ensure their full con-sideration (NAE-NRC, 1970).

One process burns the coal in a fluidized bed of lime-stone particles which react with the sulfur. A portion of the bed is continuously removed and replaced with fresh limestone. In England the National Coal Board is studying atmospheric and pressurized systems, and Esso Research Ltd. has been developing a two-stage system in which high-sulfur residual fuel oil is burned without fuel sulfur going into the gas. In the United States, the Office of Coal Research and NAPCA (now APCO) have supported research at the firm of Pope, Evans and Robbins, including pilot plant work on the Pope-Bishop fluidized-bed boiler and studies of the applicability of fluidized-bed combustion to industrial boiler systems. The Nation-al Coal Board projects that scale-up to a 20 to 30 mW atmospheric pilot unit can be completed by 1972. The Pope-Bishop boiler could be ready for market by 1972 at a capacity of 300,000 lbs/hr, factory-assembled for rail shipment. If it is successful, boiler manufacturers should be able to extrapolate the concept to very large utility boilers before 1980 (Ehrlich, 1970). The propo-nents of fluidized-bed boilers claim, in addition to sul-fur-dioxide control, significant economic advantages over conventional pulverized-coal systems. The fluidized-bed system may offer an additional advantage with respect to NO pollution control because of its ability to operate at lower temperatures than those existing in pulverized coal flames.

The second process prevents SO_2 emission by the sub-

merged partial combustion of coal in a bath of molten
iron. The resulting carbon monoxide is burned above the
melt, and the sulfur is removed with the ash slag and
subsequently recovered. Research on this process is being supported by APCO at the firm of Black, Sivalls, and
Bryson. Feasibility studies are being conducted and a
pilot plant is being designed. The projected economics
of the process are said to compare favorably with the
lower estimates for conventional power plants employing
stack-gas cleaning.

2.2.5. Stack-Gas Cleaning

At least 25 processes for removing SO_2 from stack gases
are under development in the United States, and others
are under development in Japan and Europe. Most are at
stages of development ranging from bench-scale laboratory
projects to pilot-plant studies (10 to 25 mW). Some have
been installed in operating power plants in units up to
175 mW. Several of these processes will probably be technological successes, but the efficiencies are not yet
well established for even the most advanced ones; and
industrially proved* technology for SO_2 removal from
stack gases does not exist (NAE-NRC, 1970). There is an
urgent need for commercial demonstration of the most
promising processes to make reliable engineering and economic data available for the design of full-scale facilities needed in meeting local and regional conditions.
Present (July 1971) economic estimates based on paper
studies and, to the extent available, on pioneering utility experience in full-scale installations indicate
equipment costs ranging from $20 to $40/kW (corresponding
to capital charges of 0.6 to 1.2 mills/kWh for installa-

---

*The definition of proved industrial-scale acceptability
is satisfactory operation of a 100-MW or larger unit for
more than one year.

tion on new modern power stations with most of the plants'
economic life remaining). Actually, these installation
costs may be even higher, and this will certainly be the
case for the more difficult backfitting projects. For
older plants the capital charge can increase to several
mills/kWh. The changing scene is indicated by the fact
that the NAE panel estimates of late 1970 and APCO's estimates of early 1971 are bracketed by the capital cost
figures $4 to $40/kW. Operating costs ranging from 0.5
to 2.2 mills/kWh bracket the NAE and APCO estimates.
Discussion of the many processes (one research group has
counted 76) is beyond the scope of this report; the following list indicates some of the characteristics of several of the better-known processes:

Process Name, Developer, and Other Features	Stage of Development*
Limestone Injection-Limestone Wet Scrubbing, and Wet Scrubbing Alone. (Combustion Engineering Co.) An add-on process with large waste disposal problem.	Full-scale comm'l units to 530 MW being tested
Limestone-Dry Removal. Add-on; electrostatic precipitation.	(175 MW)
Cat-Ox Process (Monsanto). Catalytic production of sulfuric acid; stack-gas treatment at 700-900 F; best for high-S fuels; requires boiler modification.	(250 MW)
Wellman-Lord Process. Add-on; sodium sulfite scrubber. Reprocessing for contact-acid mfg.	(25 MW; comm'l scale test planned)
Esso-Babcock and Wilcox Dry Adsorbent Process. Operates on 900 F gas; re-quires boiler modification; sulfuric acid mfg.	(25 MW)

----

*Pilot or demonstration plant (size), or bench operation.

Chemico Magnesium Oxide Scrubbing   (150 MW)
Process. Scrubs with slurry; revivification at central chemical
plant; add-on.

Formate Scrubbing Process (Consoli-   Pilot studies
dation Coal Co.). Add-on; absorbent
reprocessed to feed Claus sulfur plant.

Ammonia Scrubbing. Gas cooling, wet   Japan
scrubbing, gas reheat; product is
ammon. sulfate, nitrate or phosphate;
add-on

Westvaco Char Process. Uses adsorbent-   Pilot plant
catalyst; regeneration with H_2 to
produce Claus-plant feed.

Molten Carbonate Process (Atomics   Bench-scale
International). Gas treatment at
900 F, reprocessing of melt with
H_2-CO mix to produce Claus-plant
feed; boiler modification required.

Sodium Bicarbonate Adsorption Process   Pilot-level
(Dow Chemical). Add-on; removes fly-
ash as well.

Modified Claus Process (Consolidation   Bench
Coal Co.). Uses S + H_2 to make H_2S
to add to SO_2 in stack gas to make
S by Claus-process; add-on.

Catalytic Chamber Process (Tyco   Bench
Laboratories). A modification of
old lead chamber acid process; add-on.

Ionics/Stone and Webster Process.   Pilot-level
Scrubber and electrolytic cell sys-
tem to recover SO_2 for acid mfg;
add-on.

Alkalized Alumina Process. Unaccep-   Dropped
table because of attrition and
pore-blockage problems.

Sulfacid Process (Lurgi): wet char   Germany
sorption; add-on.

2.2.6. Stack Gas Dispersion

Good stack-gas dispersion, such as that afforded by tall
stacks and favorable weather conditions, allows higher
SO_2-emission limits to satisfy atmospheric standards,
thereby reducing the extent to which other sulfur dioxide
abatement techniques must be applied. Thus flexible emission regulations that give proper credit for good dispersion would permit significant cost reductions in SO_2-pol-
lution abatement. A bibliography on plume rise calculations, stack height, diffusion of stack gases and mathematical diffusion models is given by NAPCA (1969b).

Stack-gas dispersion will continue to be important
even when efficient SO_2-control techniques are available.
When a 1%-sulfur fuel is being burned to produce 800 ppm
in the exit stack gas, the dilution ratio in the stack
plume at ground level must be 27,000 even where no other
nearby stack is emitting SO_2, if the ground-level requirement of 0.03 ppm is to be met.

2.3. Thermal Pollution Control

2.3.1. Introduction

The 1970 U.S. rate of energy consumption, 68.8×10^{15} Btu/
yr, corresponds to an energy release rate per unit surface area of about 1/600 of the solar incidence on the
48-state U.S. land area,* or about 1/18,000 of the global
solar incidence.** This number indicates that thermal
pollution is not at this time a global problem. Houghton
(1971) estimates, on the basis of studies of variation of
the solar constant, that important effects of man's additions to the earth's heat are not to be expected if those
----
*U.S. annual-average solar incidence is about 1400 Btu/
ft^2day (Section 7.2).

**On the assumption that 1/4 of the energy toward the
earth from the sun reaches the earth's surface.

additions amount to less than one percent of the solar constant (this is not to say there will be effects when that
limit is reached). A calculation of the U.S. and non-
U.S. growth in energy consumption,at those projected
rates to 2000 A.D. which have come out of a study by
Darmstadter and associates (1971), coupled with the assumption of subsequent world growth at 4% per year, leads
to the conclusion that more than one hundred years will
elapse before man's nuclear and fossil-fuel combustion
activities will add 1% to the solar input.

Local thermal pollution, however, arising mainly from
waste heat discharged from central power plants into
streams, lakes, and coastal waters, is already a considerable problem in some areas. (Local thermal air pollution is and will continue to be of little concern because
of natural convection in the atmosphere). A brief and in
many ways inadequate summary of information bearing on
local thermal pollution and the cost of its abatement is
presented in this section.

The 1968 report of the National Technical Advisory
Committee on Water Quality Standards recommended water
quality standards in five general areas of water use.
Regarding fish and other aquatic life, the committee recommended that heat should not be added during any month
of the year that would raise the water temperature more
than 5 F for streams with warm-water fisheries, or 3 F
for the epilimnion of lakes and reservoirs. Estuarine
temperatures should not be raised more than 4 F during
fall, winter and spring, or by more than 1.5 F during
summer.

The maximum allowable temperature and the maximum allowable change in temperature have been established by
most states as 68 F and 0-5 F, and 83-93 F and 4-5 F,
for streams with cold and warm-water fisheries, respectively. The limits of mixing zones are defined by only

some states. The policy of many states prevents degradation of waters having a quality that exceeds the accepted
standard.

The most common scheme for the control of thermal pollution in power generation is to use, instead of once-
through cooling water in the condenser, recycled water
which loses heat to air in a cooling pond or cooling tower. Thus the heat is transferred directly to the atmosphere without thermal alteration of natural waters, a
solution which has little large-scale impact since the
ultimate heating of the atmosphere is unaffected by mode
of energy dissipation into it.

About 16% of the installed steam-electric power plants
larger than 500 MW now use cooling towers, and 8.5% use
cooling ponds. The Federal Power Commission (Jimeson and
Adkins, 1971) estimates that as many as 140, and possibly
as many as 300, new stations built by the year 1990 will
require cooling towers, and 50 to 130 will require cooling ponds. Legislation has been proposed which would require auxiliary cooling for plants operating on a sea-
coast. Recent hearings into thermal discharges into Lake
Michigan call for cooling towers at all new power stations, and there are demands that older power plants and
industrial facilities also be similarly equipped.

2.3.2. Heat Dissipation in Power Plant Operation

Modern fossil-fuel-burning power plants have a thermal
efficiency of about 40% and lose in the stack gas and in
the plant about 10% and 2% of the total energy input.
Therefore, the rate of heat dissipation to the cooling
water of a 1000 MW plant is about 1200 MW, which requires
a cooling-water flow rate of about 900 ft^3/sec if the
temperature rise of the water is 20 F.* In comparison

----

*This figure represents the typical range 10-30 F.

with this plant, a 1000 MW light water nuclear reactor
has an efficiency of about 33%, loses in the plant about
1% of the heat input, and therefore rejects in the condenser about 2000 MW, thus requiring a cooling-water flow
rate of about 1500 ft^3/sec. Because of the larger thermal waste from nuclear plants the Federal Power Commission, which expects an increase in nuclear power from the
1970 value of 1.9% of the total to 22.1% in 1980 and
39.7% in 1990, projects a considerably faster fractional
increase in the national rate of waste-heat production
than the 10-year doubling time usually assumed for power
generation. Power-station waste heat is projected to increase from the 1970 value of 5.3 x 10^{15} Btu/yr to 12.8 x
10^{15} in 1980 and 28.4 x 10^{15} in 1990. The projected
rate of consumption of cooling water, assuming that consumption (evaporation) continues to be about 1% of the
condenser flow in once-through cooling systems and about
2% in wet cooling towers, is 4300 to 6600 ft^3/sec and
13,800 to 14,700 ft^3/sec in 1980 and 1990. These figures
are to be compared with the 1970 consumption rate of
1400 ft^3/sec. The 1990 evaporation rate of 14,000 ft^3/
sec is 0.8% of the national average runoff. This discharge of vapor is of such magnitude that extensive fed-
erally-sponsored studies of its microclimatological effects are certainly warranted.

2.3.3. Types of Cooling Systems for Power Plants
There are three types of cooling systems for large power
plants: Once-through systems utilizing large bodies of
water (rivers, lakes or oceans) as heat sinks; evaporative cooling systems (mechanical-draft or natural-draft
cooling towers, or cooling ponds) in which heat is dissipated to the atmosphere by recirculating cooling water
which partly evaporates; dry exchange systems (mechanical
or natural draft) in which a closed coolant loop trans-

fers heat to the atmosphere with no evaporation. Choice
among these depends upon water availability, land costs,
meteorological conditions, capacity penalties imposed on
the system by increases in condenser backpressure, legis-
lative restriction, and other factors. If a cooling pond
is used, the required surface area is about 1 acre per
megawatt. Cooling towers, which are of increasing importance, are described briefly below.

Natural-Draft Wet Towers. There are three types: spray,
counterflow-packed and crossflow-packed. The chimney-
like hyperboloid towers, more common in Europe than in
the United States, are usually built of thin-shell rein-
forced concrete. The height is 300 to 500 ft, the base
diameter is in some cases as large as 400 ft, and the top
diameter is of the order 2/3 that of the base. The station power per tower is usually 250 to 900 MW, but towers
with 1100 MW capacity are available.

Spray towers contain no packing. Water sprayed near
the top flows downward against the air which enters
around the periphery at the base of the tower. In counterflow-packed towers the water flows downward over a
packing distributed over the cross section. The crossflow type has a packing around the periphery at the base,
across which the entering air flows horizontally and over
which the water flows down. In all three types a basin
beneath the tower serves as a reservoir for the circulating cooling water. These towers are best suited to regions with high humidity, and they are desirable in popu-
lated areas because the height of the exhaust point helps
prevent fogging and drizzle or icing.

Mechanical-Draft Wet Towers. This type, which may be
counterflow or crossflow, is 20 to 60 ft in height and
composed of multiple units or cells 35 to 75 ft on a side,
each with its own fan in the top and its own water distribution system. A 1000-MW station uses 32 to 36 cells.

The low height makes these towers aesthetically more acceptable, but less desirable than the taller natural-
draft towers if local conditions are such that recirculation of the exhaust air into the air intake or precipita-
tion in the vicinity of the plant are problems. On the
other hand, these towers offer more flexibility of control and may be used in areas where climatic conditions
(such as low humidity and high air temperature) make natural-draft towers poor operators.

Mechanical- and Natural-Draft Dry Towers. There are two
types of systems, indirect, in which turbine exhaust
steam is condensed by direct contact with jets of return
water from the cooling tower, with some of the condensate
going to the feedwater circuit but most returning to the
cooling tower, and direct, in which the turbine exhaust
steam passes directly into the coils of the cooling tower.
The principal difference between the two types is that
the tower coil must handle a much greater volume of low-
pressure steam in the direct system than of circulating
water in the indirect system. The direct system is generally limited to turbine-generator sizes of 200 to 300
kW, whereas the indirect type can be constructed for generating unit sizes up to 1000 MW and larger. Dry cooling
systems are characterized by negligible water consumption
and freedom from scaling, corrosion and fouling of heat-
exchange surfaces.

2.3.4. Costs of Thermal Pollution Control in Power Plant
Operation

The effect of cooling towers on the cost of electricity
has been studied by several authors. Each has made allowance for the effects of auxiliary power requirements
for circulation, efficiency due to changed condenser
pressure, and associated changes in capital charges and
fuel consumption. The extent of disagreement among the

results is indicated by Table 2-8, which gives the incremental cost, above use of the simplest once-through cooling system such as a river, that is associated with various cooling schemes in application to a large power plant
(about 800 MW; varies among authors).

The last column gives the average estimated incremental cost, mills/kWh. There is reasonable agreement that
cooling ponds will add up to 0.1 mill/kWh, wet towers 0.1
to 0.2 mill. There is less agreement on dry towers;
Woodson and ORNL think they will add up to 1 mill more
than wet towers, Rossie, 0.5 mill, and Oleson & Boyle,
0.6 mill. Natural-draft towers will add more than mechanical-draft towers, according to Woodson and ORNL; but
Rossie and Oleson & Boyle apparently consider the difference small. The natural-draft tower appears to be increasing in favor.

Space and time prevent a review of the extensive research going on in this area, of the conflict between
those ecologists who fear for the changes produced in
aquatic life by increased temperature and those who be-
lieve that some of these changes can be beneficial especially in marine heat dissipation, and of suggestions of
waste-heat utilization such as in aquaculture and agri-
culture.

2.4. Energy Storage

In the various energy systems for power production, thermal processing and space heating, the provision of one or
more points of energy storage in the system flowsheet can
improve the performance or reduce the cost or both. Energy storage can take the forms of fuel storage, such as a
coal pile, the heating oil in a householder's oil tank,
the gasoline in his car; electrochemical energy storage,
as in storage batteries; material storage for later work
production, as in pumped water; thermal storage, as for

Table 2-8. Incremental* Cost of Thermal Pollution Control in Power Plant Operation**

Type of Cooling System	Incremental* Capital Cost Associated with Cooling, $/kW				Incremental* Power Costs due to Fuel, Operating, and Capital Charges, mills/kWh				
	RDW	ORNL	JPR	J & A	RDW	ORNL	JPR	O & B	Average
FOSSIL FUEL									
Cooling Pond	2.3	3-5		2-3	0.06			0.1	0.08±0.02
Wet tower Mech. Draft	2.5	7-15		3-5	0.08	0.04	x	0.2	0.10±0.07
Wet tower Nat. Draft	5.5			4-6	0.14	0.20			0.18±0.03
Dry tower Mech. Draft	15.7	35-50	17	16-17	0.68	0.95	x+0.48	0.8	0.81±0.11
Dry tower Nat. Draft	33.3		20	18-21	0.98	1.19			0.99±0.16
NUCLEAR FUEL									
Cooling Pond	3.0	4-7		3-4	0.06			0.1	0.08±0.02
Wet tower Mech. Draft	4.2	10-12		5-6	0.09	0.02		0.2	0.10±0.08
Wet tower Nat. Draft	9.5			6-8	0.20	0.23			0.21±0.01
Dry tower Mech. Draft	27.7	50-60	23	23	0.86	1.15		0.8	0.93±0.15
Dry tower Nat. Draft	54.8		27	25-27	1.40	1.32			1.17±0.27

* The base case is cooling by once-through water flow, as from a river.

**Costs from literature identified in column headings as follows: RDW = Woodson (1971);
ORNL = Oak Ridge National Laboratory (1970); JPR = Rossie (1971); J & A = Jimson and
Adkins (1971); O & B = Oleson and Boyle (1971).

electric-load peak-shaving associated with space heating,
and mechanical storage, as in flywheels or gas pressure
vessels. These will be considered briefly.

2.4.1. Energy Storage as Fuel

The coal piles found at power plants, coke ovens, and
coal-using processing operations represent an energy
storage which good management tries to avoid because of
the oxidative deterioration of coal, the cost of capital
tieup, and the space requirement. The unit train and the
integral train for coal are important contributors to
lessened storage and transportation costs.

Fuel oil storage in oil-heated domestic dwellings
could represent a significant means of easing the seasonal storage problem of the petroleum industry and of reducing the cost of fuel-oil distribution. Not until the
oil industry shares some of the gain with the householder,
however, will the latter develop much enthusiasm for investment in a 1000 gallon tank instead of one holding 265
gallons. A systems analysis of this problem appears to
be warranted.

The ability to store energy compactly as gasoline is
well known. The fact that a 200-pound load can be **lifted**
from sea level to the top of Mount Everest with one and
one-half pints of gasoline burned in an engine of 28%
efficiency is of course a major reason why we don't yet
have many electric autos. Although there are other fuels
of higher energy storage capacity per unit mass--liquid
hydrogen, aluminum or magnesium powder are examples--
there is negligible expectation of competing with the
combination of high value and low cost that is gasoline.

2.4.2. Electrochemical Energy Storage

Storage batteries deliver electric power by electrochemical reactions occurring at two electrodes immersed in an

active solution--the electrolyte; and they accept and
store electric energy as chemical energy by a reversal of
the chemical reactions. The voltage of a single battery
cell is determined by its chemical makeup, not its size;
and desired higher voltages are obtained by connecting
cells in series. Withdrawal of electric power causes the
voltage to drop somewhat, due to internal resistance of
the battery. There are two important and interdependent
characteristics of a storage battery, its specific energy
storage, expressed as watt hours per pound of total
weight, and its specific power, or rate at which it can
deliver energy; and the first of these goes down when the
second goes up because of the battery's internal resistance. Two other obviously important characteristics are
cost and life, the latter measured by the number of cycles or recharges possible. The status of storage-bat-
tery development has been well covered by a Panel on Electrically Powered Vehicles, Subpanel on Energy Conversion
and Storage Systems, reporting to the U.S. Department of
Commerce (1967). The comments that follow are based
largely on that report.

Commercially Available Batteries. Figure 2-9 presents
the ratings of both commercially available and a few newer systems, expressed as specific power (Watts/lb) versus
specific energy (Watt hrs/lb). Comments follow.

Lead-Acid. This is the familiar automobile starting battery, reliable, efficient, rugged, and moderately priced.
Starter batteries have a life of about 200 cycles and an
energy delivery capacity, at low output, of 15 Watt hours
per pound. Traction batteries have a longer life--1000
cycles--but lower storage capacity, about 10 Watt hrs/lb.
Lead-cobalt batteries add cobalt salt to the positive
plate and increase the energy delivery from 15 to 18 Watt
hrs/lb.

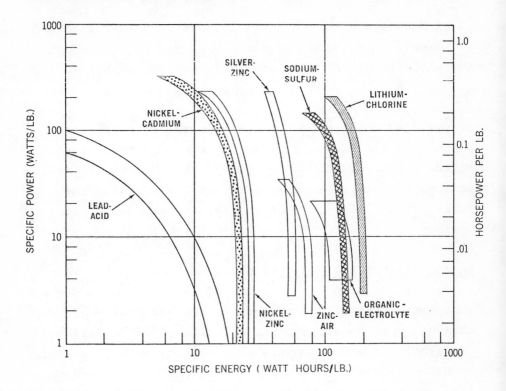

Figure 2-9. Specific Power Versus Specific Energy for
Batteries (U.S. Department of Commerce, 1967)

Nickel-Iron. These have been known for many years as
Edison cells. They are rugged and of long life--3000
cycles--but their energy storage and power output are below lead-acid.

Nickel-Cadmium. These have excellent life (2000 cycles),
a limiting energy density of 15-22 watt hrs/lb, and a

power density up to 160 watts/lb, with 300 to 350 expected from construction changes. They are expensive, and
the world supply of cadmium is limited and linked to zinc
production.

Silver-Zinc. The performance is superior, except for
short life (100-200 cycles). The use of 5 to 6 pounds of
silver per kilowatt hour of stored energy makes this battery impractical for widespread use. Silver-cadmium offers better cycle life but is also too expensive for general use.

Relatively New Battery Systems.

Zinc-Air. Initially a primary (nonrechargeable) battery,
this has been developed into a rechargeable battery by
circulating the electrolyte with a pump, with a resulting
high recharging rate and a specific energy delivery of
60-80 watt hrs/lb. A 14-kWhr prototype costs a little
less than lead-acid batteries, i.e., $50-60/kWhr. Opera-
tion is at ambient temperature. Static electrolyte batteries are being built with a projected energy capacity
of over 100 watt hrs/lb.

Organic Electrolyte. These can use lithium or sodium as
one electrode because of the absence of water, with which
those metals would react explosively. Promising combinations use a lithium anode and copper or nickel chloride
or fluoride as cathode, with propylene carbonate as the
electrolyte base. Power is not high--20 to 30 watts per
pound--but the specific energy has reached 100 and is
projected to reach 150 watt hrs/lb. Cycle life is a
problem.

Sodium-Sulfur. All batteries discussed up to this point
have been capable of operation at normal temperatures.

This one, developed by Ford Motor Company, operates at
250-350 C.   Liquid sodium and sulfur combine to form
Na_2S_3 in a solid "electrolyte" of beta-alumina.   Cost estimates are in the range of $20 to $40 per kilowatt hour,
both power and energy density are attractively high, and
the permitted charging rate is comparable to the maximum
discharge rate.   A two-kilowatt-hour module should have
been constructed by now.

Fused Salt Electrolytes.   The lithium-chlorine battery
with molten lithium chloride electrolyte, developed by
General Motors Corporation, is in this class.   The battery operates at 600-700 C, and molten lithium, molten
LiCl and gaseous chlorine must be accommodated for long
periods of time with appropriate electrical insulation
and no leakage.   Specific power and specific energy are
high (Figure 2-9) and permitted recharging rate is high;
but so is 700 C.

Nickel-Zinc.   Substitution of zinc for cadmium in the
nickel-cadmium cell previously described lightens and
cheapens the battery.   Energies in the range of 25 to 30
watt hours per pound are considered possible, but recharging the zinc electrode presents problems.

The importance of cheap good storage batteries is so
clear that there is no problem of stimulating research in
this area.   Each year brings more patents on new combinations of chemicals.   Opinion is divided on whether high-
temperature systems, with their superiority in performance and their obvious containment and safety problems,
will ultimately be developed to the point of acceptability.

2.4.3.   Energy Storage in Pumped Water
There is little need to comment here on an old idea, be-

yond recording the evidence of growth of its application.
The opportunities for storing energy by pumping water to
a high storage reservoir for later use are limited, but
installations are growing in number. Table 2-9 presents
the record of additions to pumped storage in 1969 and of
planned additions as of December of that year.

For comparison, hydroelectric power stations and total
generating capacity are included. The table indicates a
planned increase in pumped storage from none to about two
percent of the nation's total generating capacity.

2.4.4. Thermal Storage

Energy can be stored by the heating, melting or vaporizing of a material; and the energy becomes available as
heat when the process is reversed. Storage by causing a
material to rise in temperature is called sensible-heat
storage; its effectiveness depends on the specific heat
of the material (in English units, the Btu's required to

Table 2-9. Annual Hydroelectric Power Additions, Made
            and Planned, 10^3 MW

	Hydro	Pumped Storage	Total Genera- ting Capacity
1969	2.3	0.8	24.1
1970	2.1	0.2	32.1
1971	1.1	1.1	32.7
1972	0.5	0.9	37.9
Total planned, as of Dec. 1969	10.8	10.6	206.9
Total in exist- ence, Dec. 1969	52.7	-	312.6

raise 1 pound 1 degree F) and, if volume is important, on
the density of the storage material as well. Storage by
phase-change--the transition from solid to liquid or from
liquid to vapor--is another mode of heat storage, known
as latent-heat storage, in which no temperature change is
involved (though both sensible- and latent-heat storage
may occur in the same material, as when a solid is heated,
then melted, then raised further in temperature). Heat
of fusion generally is equivalent in magnitude to the
product of specific heat by a rather large temperature
change, and its exploitation is therefore an attractive
concept in heat storage. Table 2-10 presents data on a
number of materials which could be used for sensible or
latent heat storage.

   Table 2-10 suggests the following generalizations:
1.
For sensible-heat storage to temperatures below 200 F
water is outstanding; 157 cu ft of water rising 100 F
will store the energy released by burning 10 gallons of
fuel oil at 70% efficiency (one-day winter fuel consumption of a modest domestic dwelling).
2.
For sensible-heat storage with air as the energy-trans-
port mechanism, gravel or crushed stone in a bin has the
advantage of providing a large, cheap heat-transfer surface; the bin volume will, however, be two and one-half
times the volume of a water tank that is heated over the
same temperature interval.
3.
For latent-heat storage in the vicinity of room temperature no good material has been found. Glauber's salt
($Na_2SO_4 \cdot 10 \ H_2O$) has been suggested, but its phase-change
at 97 F is not a complete melting but a separation into
three phases of different density and composition; the

Table 2-10.   Heat-Storage Materials

Sensible-Heat Storage	Sp. Ht., Btu/lb F	True Density, lb/ft^3	Heat Capacity, Btu/ft^3 F	
			No Voids	30% Voids
Water	1.00	62	62	–
Scrap Iron	0.12*	490	59	41
Magnetite (Fe$_3$O$_4$)	0.18*	320	57	40
Scrap Aluminum	0.23*	170	39	27
Concrete	0.27	140+	38	26
Stone	0.21	170+	36	25
Brick	0.20	140	28	20
Sodium (to 208 F)	0.23	59	14	–

Latent-Heat Storage	Melting Point, F	Density, lb/cu ft	Heat of Fusion, Btu/lb	Btu/cu ft
Calcium Chloride hexahydrate	84-102	102	75	7,900
Sodium Carbonate decahydrate	90-97	90	115	10,400
Glauber's Salt	90	92	105	9,700
Sodium Metal	208	59	42	2,500
Ferric Chloride	580	181	114	20,600
Sodium Hydroxide	612	133	90	12,000
Lithium Nitrate	482	149	158	23,500
Hypophosphoric Acid	131	94?	92	8,700
Lithium Hydride	1260	51	1800	92,000

*Over interval 77 F to 600 F.

phase-change is in consequence not rapidly enough rever-
sible to make 24-hour storage feasible.* If melting is to
occur near the desired space temperature (say, between 90
and 120 F), there are limitations on materials. Organic
materials generally have such weak crystal lattices that
their heat of fusion is too low to be interesting. Inorganic salts generally melt at higher temperatures unless
they are hydrates; and then the common occurrence is separation into an anhydrous solid residue and a dilute solution rather than true melting. What is desired, then, is
an inorganic hydrate which is cheap and which has what is
known as a congruent melting point, i.e., the water content of the melted and unmelted parts is the same. Melting or solidification at the interface between phases is
not then accompanied by dependence on molecular diffusion,
which is slow, and rapid melting or solidification can
occur as heat is added or removed.

4.

Heat storage over a large temperature interval. Iron ore
(magnetite) has the unusual property of exhibiting over a
temperature range to above 1000 F a volumetric heat capacity about the same as water at low temperature. Reduced
to 70% because of voids between the ore lumps, the capacity is still impressively high. A bin of 37 cu ft volume, with the ore rising 700 F, could store the energy of
the example in item 1.

5.

For storage over a large temperature interval concrete is
not as efficient, volumetrically, as magnetite, but it
has obvious advantages. The use of large insulated con-
crete-block storage units is common in England to iron
out the electric home-heating load.

6.

High-temperature latent-heat storage has some striking

---
*It is claimed that thickening and nucleating agents have
now solved this problem.

possibilities. Ferric chloride melts at 580 F with a
heat of fusion of 20,600 Btu/cu ft; lithium nitrate at
482 F with 23,500; hypophosphoric acid at 131 F with
8,700. The first of these could have economically signi-
cant use in smoothing electric space-heating loads.

The Economics of Thermal Storage. It is probably fair to
say that storage has many times been suggested as a solution to an energy problem when it would in fact increase
the cost because of its poor use factor. Two examples
will be given, one almost trivial but familiar, and one
on which much time and engineering talent have been spent.

Oil refineries must infrequently get rid of a rather
large volume of hydrocarbon gases in a short time, in
connection with changeover from one mode of operation to
another, change of feed stock, or some other transient in
the operation. The discharge, however, occurs so infrequently and so irregularly that the provision of storage--
a gasholder of the type common in cities in pre-natural
gas days--would incur a capital charge many times the
value of the fuel saved. So the gas is burned, with occasional protest by an observer to the effect that the
price of gasoline is being raised by the prodigal waste
of good fuel gas.

Solar house heating involves the warming of a fluid,
water or air, by passage through flat-plate collectors on
the house roof, and the warming of the house interior
with that fluid. The thermal input from sun to roof varies with time in an entirely different pattern from the
thermal needs of the house; and thermal storage is necessary. For a given house size and seasonal weather pattern the total heat load versus time is established.
The optimization of choice of roof collector area and
heat storage volume is then in principle a complex problem in two-variable optimization. One would tend to as-

sume that, since the winter sun is rather undependable, there must be provision for at least several days' energy storage to keep the house warm for a succession of dull days. But this implies provision of a roof area sufficiently large to supply January's day-by-day needs and in addition "fill" the large heat-storage unit. A roof designed for coping with January-February, however, would be grossly overdesigned for the remainder of the heating season, and there would be no use for much of the energy collected, that is, no income to balance much of the fixed capital charge on the roof. The conclusion is thus reached that an auxiliary heat source--a gas- or oil-fired furnace (electric rates on a house not dependent exclusively on electric heat would be prohibitive)--would be necessary and that the concept of three-day storage for January is out. Once an investment is made in auxiliary heating, the concept of two-day storage also drops out, since the number of times it is used per year is too small to pay the capital charge on it.* The conclusion is thus reached that the only heat storage making sense in a solar-heated--or, realistically, a partially solar-heated--house is overnight storage; when the sun doesn't come out, fuel is usually burned. This conclusion is not changed until a method of heat storage is found which is so cheap that only infrequent use of it would pay its fixed charges. If such a storage system were ever found, its use in solar house heating would be one of the least among its many applications.

2.4.5. Mechanical Storage
The possibility of storing energy in a flywheel for opera-

----

*Two-day storage could be justified in special cases, particularly when the climate pattern is such that heating needs are distributed relatively uniformly through the season (Tybout and Löf, 1970).

tion of an automobile has been suggested. With the last
stages of steam turbines operating with top speeds of
2000 ft/sec, it is reasonable to assume that an automobile flywheel could be safely designed for 1000 ft/sec
rim speed. If 120 pounds were located at the rim, the
kinetic energy stored would be $(120)(1000)^2/(2)(32.16)$
foot pounds or 0.71 kWh. This is the energy storage capacity of the usual automobile storage battery. Time has
not permitted proper examination of the development status of this suggestion. Clearly, the development of new
materials such as fibre-reinforced flywheel rims could
aid progress. Doubling the rim speed would cut the mass
to one quarter.

References

Cabinet Task Force on Oil Import Control, 1970. The Oil
Import Question, Washington, D.C., U.S. Gov't. Printing
Office, p. 247.

Chemical and Engineering News, 1969. "Large Supertankers
No Longer Offer Economic Purchase Price," 47, May 26,
1969, p. 26.

Committee on Interior and Insular Affairs, 1962. National Fuels & Energy Study Group Report, U.S. 87th Congress,
2nd Session, Senate.

Darmstadter, J., and associates, 1971. Energy in the
World Economy, to be published by the Johns Hopkins Press,
late 1971.

Department of the Interior, 1962. Report to the Panel on
Civilian Technology on Coal Slurry Pipelines, Washington,
D.C., U.S. Gov't. Printing Office.

Dillard, J. K., 1965. "Transmission above 700 kV Hits
Economic Roadblock," Electric Light and Power, 43,
February, 1965, pp. 44-49.

Energy R&D and National Progress, 1964, Washington, D.C.,
U.S. Gov't Printing Office.

Environmental Science and Technology, 1971. "National
Air Quality Standards Finalized," 5, No. 6, p. 503.

Ehrlich, S., 1970. Environmental Science and Technology
4, p. 396.

Federal Power Commission, 1964. National Power Survey,
Washington, D.C., U.S. Gov't. Printing Office.

Federal Power Commission, 1970. National Power Survey,
Washington, D.C., U.S. Gov't. Printing Office.

Foster, B., 1970. "Projected Costs of Alternate Sources
of Gas," Institute of Gas Technology, Illinois Institute
of Technology, Chicago.

Gas Engineers Handbook, 1965. The Industrial Press, N.Y.

Graneau, P., 1970. "Economics of Underground Transmission with Cryogenic Cables," IEEE Transactions on Power
Systems and Apparatus, PAS-89, January, 1970, pp. 1-7.

Houghton, H. G., 1971. Department of Meteorology, Massachusetts Institute of Technology, personal communication,
August, 1971.

Hunsaker, B., 1966. "How Heating Value Affects Gas Trans-
missibility," 42, February, 1966, pp. 51-56.

Institute of Gas Technology, 1969. "LNG: A Sulfur-Free
Fuel for Power Generation, Clearinghouse for Federal Scientific and Technical Information, Washington, D.C., May,
1969, pp. 2-242.

Jimeson, R. M. and Adkins, G. G., 1971. "Waste Heat Disposal in Power Plants," Chemical Engineering Progress
67, No. 7, p. 64.

Minnich, S. and Fox, G., 1969. "Cryogenic Power Transmission," Low Temperatures and Electric Power, International Institute of Refrigeration, Paris.

NAE-NRC, 1970. "Abatement of Sulfur Dioxide Emissions
from Stationary Combustion .Sources," COPAC-2, National
Academy of Engineering - National Research Council.
Washington, D.C.

NAPCA, 1969a. "Air Quality Criteria for Sulfur Oxides,"
U.S. Dept. of Health, Education and Welfare, National Air
Pollution Control Administration, NAPCA Publication AP-50,
Washington, D.C., January, 1969.

NAPCA, 1969b. "Control Techniques for Sulfur Oxide and
Air Pollutants," U.S. Dept. of Health, Education, and
Welfare, National Air Pollution Control Administration,

Washington, D.C., January, 1969, pp. 102-105.

Oak Ridge National Laboratory, 1970. "Interdisciplinary Research Relevant to Problems of Our Society," Progress Report to NSF on Summer Study of 1970, Report ORNL-4632, December, 1970.

Ode, W. H., 1963. "Coal Analysis and Mineral Matter," Chapter 5 in Chemistry of Coal Utilization, Supplementary Volume (H. H. Lowry, Ed.), John Wiley and Sons, p. 215.

Oil and Gas Journal, 1966 64, July 4, 1966, pp. 84-85.

Oleson, K. A. and Boyle, R. R., 1971. "How to Cool Steam-Electric Power Plants," Chemical Engineering Progress 67, No. 7, p. 70.

Papamarcos, J., 1970. "ERC's Underground Transmission Program: A Status Report," Power Engineering 74, February, 1970, pp. 26-30.

Rossie, J. P., 1971. "Dry-Type Cooling Systems," Chemical Engineering Progress 67, No. 7, p. 58.

SCEP, 1970. Man's Impact on the Global Environment--Assessment and Recommendations for Action, Report of the Study of Critical Environmental Problems, M.I.T. Press, Cambridge, Mass., p. 161.

Tsaros, C. L., 1971. Institute of Gas Technology presentation before the Coal Gasification Panel of the National Academy of Engineering, June, 1971.

Tybout, R. A. and Löf, G.O.G., 1970. "Solar House Heating," Natural Resources J. 10, No. 2, pp. 268-326.

U.S. Dept. of Commerce, 1967. "The Automobile and Air Pollution: A Program for Progress," Part II, Subpanel Reports to the Panel on Electrically Powered Vehicles, U.S. Gov't. Printing Office, December, 1967.

West, J. R., 1970. "In Support of CGI Cable," Power Engineering 74, March, 1970, pp. 42-43.

Woodson, R. D., 1969. "Cooling Towers for Large Steam-Electric Generating Units," Chapter in Electric Power and Thermal Discharges (M. Eisenbad and G. Gleason, Eds.), Gordon and Breach, p. 351.

Chapter 3

FOSSIL FUEL-TO-FUEL CONVERSION

3.1. Pipeline Gas from Coal

3.1.1. Some Guiding Principles

A knowledge of some of the principles underlying the vaious processes being developed for producing a natural-gas substitute from coal will be helpful in understanding the processes and identifying their similarities and differences. Given an interest, in connection with coal gasification, in the molecular species H_2O, H_2, CO, CO_2, and CH_4 and their reaction with solid carbon, it follows that three unique stoichiometric equations exist. Choice of the three is arbitrary, except that every species must be mentioned and no redundancy must appear. The three chosen appear below together with: (i) their enthalpy

Reaction	-ΔH at 1200 K cal/g. mole	Temperature at which $K_p > 1$	$\dfrac{d\ln K_p}{d\ln T} = \dfrac{+\Delta H}{RT}$
1. $C + H_2O_{(v)} = CO + H_2$	$- 32,457$	above 947 K (1245 F)	17.2
2. $CO + H_2O_{(v)} = CO_2 + H_2$	$+ 7,838$	below 1100 K (1520 F)	$- 3.7$
3. $C + 2H_2 = CH_4$	$+ 21,854$	below 819 K (1014 F)	-12.2

change, -ΔH at 1200 K (1700 F); (ii) the temperature above or below which the equilibrium constant K_p exceeds 1, and, at that temperature; (iii) the percent increase in K_p per percent increase in temperature (JANAF Thermochemical Tables), with no implication concerning kinetic mechanisms. The thermodynamic tendency of one or a combination of these to go strongly in one direction is sometimes mistakenly identified with the mechanism of the process.

If oxygen is introduced into the system to change the thermal balance, one additional relation is needed, conventionally,

4. $C + 1/2\ O_2 \rightarrow CO + 26,637$ cal.

At all temperatures and atomic compositions of interest here, the thermodynamic tendency of Equation 4 to go to the right is so great that no equilibrium constant is needed, and O_2 may be considered absent from all products.

Guidance as to the significance of these equations in pointing to ways of making methane from carbon and steam comes from considering the effect of temperature and pressure on the equilibrium composition of gases in contact with carbon. With an atomic hydrogen-oxygen feed ratio of 3, 2 and 1, corresponding to feeds of $H_2O + 1/2\ H_2$, H_2O, and $H_2O + 1/2\ O_2$, use of the K_p's of reactions 1-3 yields the results presented in Figure 3-1. From this figure it is clear that at a sufficiently high temperature (above 1500 F at 1 atm, 2000 F at 20 atm, still higher at 68 atm), the overall reaction tends to be the highly endothermic reaction

(1) $C + H_2O_{(v)} \rightarrow CO + H_2 - 32,457$ cal

whereas at a temperature below about 500 F the reaction tends to be the almost thermally neutral one

(5) $= \dfrac{(1) + (2) + (3)}{2}$: $C + H_2O_{(v)} \rightarrow \frac{1}{2}\ CH_4 + \frac{1}{2}\ CO_2 - 1345$ cal

At 1500 F and 68 atm with a pure steam feed the stoi-chiometric equation to give the results shown in Figure 3-1   (ca. 15 CH_4, 19 CO_2, 16 CO, 24 H_2, 26 H_2O) is approximately

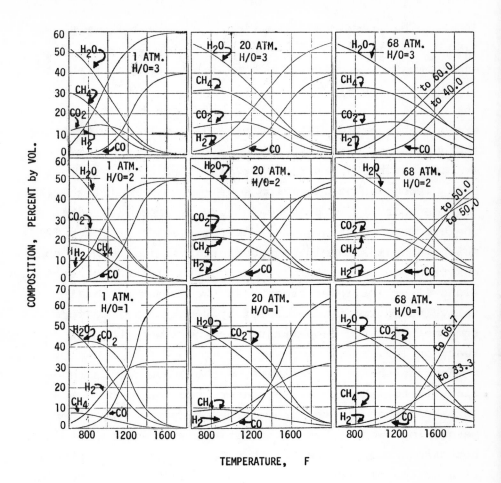

Figure 3-1. Equilibrium Composition of Carbon-Oxygen-Steam Systems at Pressure of 1, 20, and 68 Atmospheres and H/O of 1, 2, and 3

43 x (reaction 1) + 23 x (reaction 2) + 19 x (reaction 3) + 34 H_2O unreacted.

Although the low-temperature reaction (5) is thermally and composition-wise ideal, it is of course too slow at

500 F to be practical. Reaction (1) is fast above about
1750 F; and if it is to be the first step in making CH_4,
energy must be supplied, such as (a) by extreme preheat
of the entering steam (impractical), (b) by electric heating of the bed, (c) by addition of oxygen to carry out
exothermic reaction (4) simultaneously with (1) (but at
the expense of worsening the CO/H_2 ratio), or (d) by
transfer of hot inert fluid or fluidized solids to the
bed. (Proposed gasification schemes use three of these
four methods.) The presence of excess steam increases
the H_2/CO ratio through the water-gas shift reaction (2),
which maintains itself in substantial equilibrium at coal
gasification temperatures. The $CO + H_2$ formed by reactions (1) + (2), plus steam, can then react noncatalyti-
cally (but quite incompletely) by reaction (3) to form
some methane; the temperature must be lower than for
reaction (1).

An alternative to reliance on the steam-carbon reaction (1) to make $CO + H_2$ for later reaction with carbon
to make methane is the production of hydrogen by other
means, followed by its introduction, with steam, into the
hot carbon bed. There are thus three regimes of composition of the gas reacting with carbon in the upper bed, of
increasing hydrogen content: (a) the gas produced by a
steam-oxygen mixture that, in the bed layer first encountered, produces $CO + H_2$ + steam and carbon dioxide and
requires no heat, (b) the gas produced by steam alone
which, in the bed layer first encountered, produces a
higher H_2/CO ratio than (a) but requires heat (either
electrical or by transfer of inert solids), and (c) the
gas produced by addition of steam and hydrogen, with the
heat required for steam decomposition being supplied by
the methanation reaction (3). These regimes are arranged
in the order of increasing difficulty of achievement;
they are also in the order of increasing tendency for

noncatalytic methanation to occur, thereby reducing the
amount of more costly catalytic methanation necessary
later.

Additional principles that guide the choice of flowsheet are these:

1.

The volatile matter in the coal is a valuable source of
high-heating-value gaseous components, capable of destruc-
tion with accompanying deposition of carbon if the temperature of devolatilization is too high or the time of
heating too long. Consequently, flash volatilization is
in principle best, and best achieved if the coal particles are small.

2.

There is evidence that the free energy of formation of
freshly-formed carbon can be up to at least 2600 cal
above that of graphite. The coke freshly formed by flash
volatilization can thus be more active than "old" coke;
and "old" may be measured in seconds.

3.

Reaction (3) is so unfavorably affected by high temperature, with prospect of loss of CH_4 by carbon deposition,
that any other reaction which tends to increase the fraction of hydrogen present is to be encouraged. Removal of
CO_2 by absorption on calcined limestone or dolomite thus
favors reaction (3), such as by (1) + (2) + (3), i.e.,
$2 C + 2 H_2O \rightarrow CH_4 + CO_2$, or by (2) + (3), i.e., $C + CO +$
$H_2O + H_2 \rightarrow CH_4 + CO_2$.

4.

(a) When cocurrent streams of carbon and gas containing
oxygen react until solid disappears, the temperature rise
due to oxygen disappearance is continuous, and the maximum temperature occurs at the outlet. (b) When the
streams flow countercurrently the gas outlet temperature
is the same as in (a), but the solid temperature will

have reached a much higher value somewhere along the path.
(c) When the solid forms a fluidized bed, the temperature
is in rough approximation the same everywhere, and equal
to the gas-outlet temperatures of (a) and (b). Approximately, the solids in a fluidized bed undergo very rapid
mixing, but the gas is--very approximately--in plug-flow
up through the bed. If the changing composition of the
gas along its flow path is associated with an exothermic
reaction followed by an endothermic one, or vice versa,
the solids mixing that occurs in a fluidized bed is plain-
ly of high value in moving energy from where it is liberated to where it is needed. The plug-flow character of
the gas flow must not, however, be taken too literally.
Erratic bubble formation, of a scale increasing with increase in bed diameter, stands in the way of assuming
that flows in big and little systems are similar.
5.
Space permits no more than a mention of the great differences among coals and chars, affecting methane yield by
initial volatilization, tendency to soot and tar production, and reactivity judged by gasification rate.

3.1.2. Generalized Conversion Schemes

From the above considerations there emerge a large number
of possible gasification schemes, alike, however, in that
each contains, in coal-flow sequence, zones with the following primary functions:
1.
Elimination of caking.
2.
Coal devolatilization to produce some methane.
3.
Reaction of freshly devolatilized coal with H_2-CO-H_2O to
form additional methane.

4.

Preparation, from char, of the gas for reactions of
step 3.

    These four sequential operations on the coal stream
are usually but not always carried out in three distinct
zones, with two of the operations usually merged. Classification and differentiation of the various gas-making
processes will be made on the basis of the composition
and mode of production of the synthesis gas going to the
noncatalytic methanation step 3.

A.

Schemes which carry out the endothermic steam-char reac-
tion--step 4 above--directly in a unit close-coupled to
step 3. The needed energy is supplied electrically or by
a molten carbonate stream.

B.

Schemes which, again close-coupling steps 4 and 3, supply
energy to the former by adding oxygen to the steam, thereby rendering the $(H_2 + CO)/(H_2O + CO_2)$ ratio less favorable than that achieved in "A."

C.

Schemes which add H_2 to the steam before reacting the
latter with active char, thereby increasing both the
H_2/CO and the $(H_2 + CO)/(H_2O + CO_2)$ ratio in the gas going to step 3. The hydrogen is variously supplied, such
as (a) by making producer gas with char and air, using
the gas to reduce magnetic iron ore (Fe_3O_4) to Fe + FeO
and using the latter mixture to decompose steam to magnetic ore and hydrogen, or (b) by burning char with oxygen and steam to make CO and H_2 and CO_2, then adding more
steam, shift-converting catalytically, and removing CO_2
by absorption.

D.

Hybrid schemes, such as (a) use of dolomite to evolve or

capture CO_2 and supply sensible heat, and (b) production
of liquid, gas, and solid streams.

The above basis of classification will now be used to
describe the more significant and promising processes for
production of gas of pipeline quality.

3.1.3. Description of Processes

A1. Hygas-Electrothermal (Institute of Gas Technology).
The main units in this process (Figure 3-2a) are a two-
stage, fluidized-bed hydrogasifier and an electrothermal
fluidized-bed synthesis-gas generator, both operating at
1000 to 1500 psi in generally countercurrent flow of solids and gas. Caking coal (<1/8") is first made nonag-
glomerating by pretreatment (partial devolatilization)
with hot air in a fluidized bed at 1 atm and 750 F (with
off-gas not entering the product-gas stream), and is then
mixed with light oil to form a slurry which is pumped into a fluidized drying bed, operating at 600 F and 1000 to
1500 psi, where the light oil evaporates.

Coal from the drying bed passes successively through
the first stage of the gasifier where devolatilization
and partial noncatalytic methanation occur at 1300-1500 F
in the presence of hydrogen-rich gas, thence as char into
the second stage where partial gasification at 1700-1800
F occurs by reaction with steam plus hydrogen-rich gas,
thence in part as a by-product char sidestream (sometimes
0) and in part as residual char into the electrogasifier
for reaction with steam at 1800-1900 F, and finally out
as ash. Generally counter to the solids movement is the
flow of steam into the electrogasifier, the hydrogen-rich
gas from which, together with more steam, goes to the
second or bottom-stage gasifier for partial methanation,
small in amount but sufficient to supply thermal needs

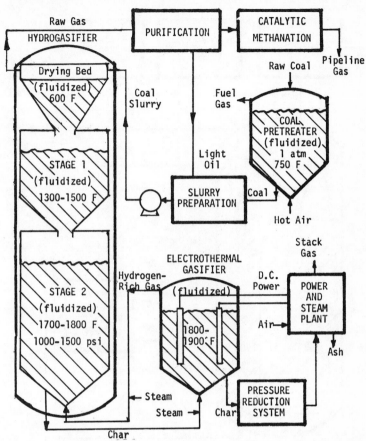

Figure 3-2a. Hygas-Electrothermal Process for Making Pipeline Gas from Coal

for the steam-carbon reaction; thence to the cooler first
stage for more methanation; thence to the drying bed and
out as product gas to the purification and catalytic men-
thanation system.

A2. Molten Carbonate (M.W. Kellogg Co.) This process
(Figure 3-2b) gasifies coal with steam at 1830 F and 420*
psia in a gasifier containing molten sodium carbonate
which serves primarily as a heat source but also as a
catalyst. Circulating melt carries char from the gasifier to a melt reheater in which the char burns with air* at
1900 F and 420 psia*, and the reheated melt returns to the
----
*Recent proposed shift to 1200 psi, and O_2 instead of air.

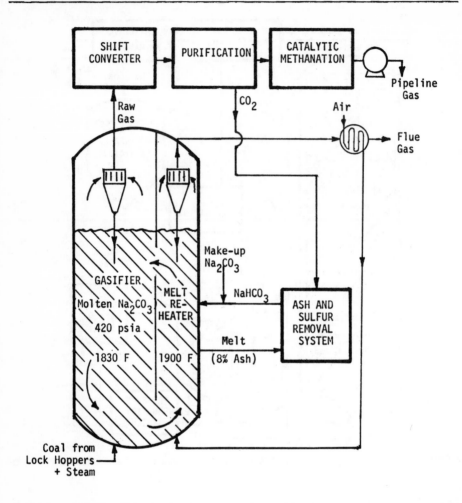

Figure 3-2b. Molten-Carbonate Process for Making Pipeline Gas from Coal

gasifier. A stream of melt is continually withdrawn and sent to an ash removal unit where the stream is quenched, dissolved, filtered to remove ash, and treated with CO_2 to drive off H_2S and precipitate sodium bicarbonate, which is returned to the gasifier. Most of the sulfur in the coal is removed by the melt as sodium sulfide and recovered as hydrogen sulfide during ash removal. Product gas from the gasifier is subject sequentially to water-gas shift conversion, removal of carbon dioxide and sulfur compounds, catalytic methanation, and compression.

B1. Bigas (Bituminous Coal Research, Inc.). This process (Figure 3-2c) uses a vertical-axis two-stage gasifier which operates at 750 to 1500 psi on either caking or
noncaking coal. Pulverized coal is injected with steam
near the bottom of the top chamber (1400-1700 F) where it
mixes with synthesis gas rising from the lower chamber
and volatilizes and partially methanates. The product
gas-unreacted char mixture leaving the top passes through
a cyclone separator from which the unreacted char stream
(94% as large as the raw coal feed stream, which indicates only a little more than 50% reaction per pass, on
the average)* is then fed tangentially into the upper
part of the lower cyclone gasification chamber where it
gasifies with oxygen and steam under slagging conditions
(2700-2800 F); the gas product is purified and catalytically methanated. The slag is water-quenched to granular
form and dropped in pressure to atmospheric by means not
yet specified.

B2. Synthane (Bureau of Mines). This process (Figure
3-2d), operating at 600 psi (with proposal to go to 1000),
gasifies pulverized caking or noncaking coal by passage
in succession through the three zones of a gasifier: (1)
a free-fall, dilute-phase coal-pretreating top section**
(750 F) in which the coal, injected with hot steam and
oxygen, is partially devolatilized, (2) a dense fluidized
bed in an expanded midsection fluidized by hot gases from

----

*The 0.94 ratio of cyclone char to gasifier raw fuel
would also satisfy such an interpretation as that 30/100
gasification occurs (mostly VCM elimination) in the raw
fuel pass through the upper section, and that 70/94 gasification occurs on passage of returned char through both
sections.

**Because of a desire to go to coarser coal feed, BuMines
has recently abandoned free-fall pretreatment in favor of
fluidized pretreating.

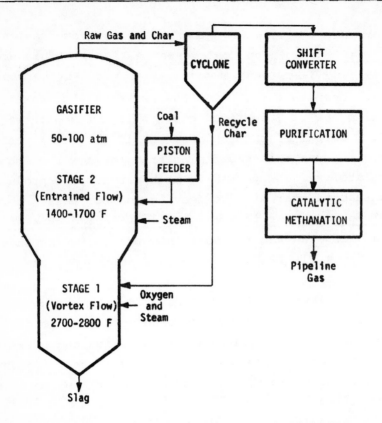

Figure 3-2c. Bigas Process for Making Pipeline Gas from Coal

below and providing the main residence time for completion of devolatilization and for uncatalyzed methanation
at 1100-1470 F and (3) a hot dilute fluidized bed in the
contracted bottom section, where entering oxygen and
steam furnish reaction heat and material for producing,
at 1750-1850 F, the synthesis-gas mixture (H_2-CO) entering section (2). Char residue is withdrawn at the bottom
of the gasifier, and the gas product leaves the system at
a point between zones (1) and (2). The product gas is
cleaned, passed through a water-gas shift converter,
scrubbed almost free of sulfur compounds and carbon dioxide, and methanated catalytically.

Cl. Hygas-Oxygen (IGT). This process (Figure 3-2e) is

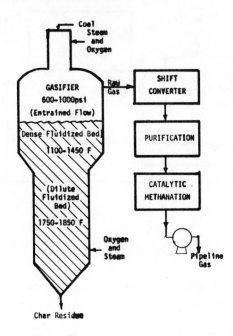

Figure 3-2d. Synthane Process for Making Pipeline Gas from Coal

similar to the Hygas process except that the electrothermal hydrogen source is replaced by a synthesis-gas generator followed by a hydrogen purification system. Synthesis gas is produced from hydrogasifier spent char, steam
and oxygen in a fluidized bed operating at the pressure
of the hydrogasifier. IGT has developed a controlled-di-
vergence feed of the oxygen-steam into the fluidized bed,
found necessary to prevent local hot spots and associated
agglomeration. The synthesis gas is shifted in composition by steam addition, catalysis, and carbon dioxide removal. The hydrogen-rich gas is mixed with steam and fed
to the second stage of the hydrogasifier. Because the
oxygen is added in a separate reactor followed by shift
conversion and CO_2 removal, considerably less is required
than in the Class B processes which add oxygen directly
to the gasifier.

An alternative to a fluidized-bed producer of synthe-

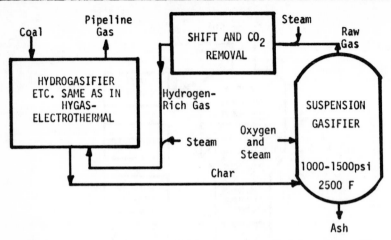

Figure 3-2e. Hygas-Oxygen Process for Making Pipeline Gas
from Coal

sis gas is the Texaco version, in which the char enters
in entrained flow with steam-oxygen and the generator is
run at 2500 F.

C2. Steam-Iron (IGT-Fuel Gas Associates). This process
(Figure 3-2f) is a modification of the Hygas process in
which the electrothermal hydrogen source is replaced by
a steam-iron system (see several pages back) operating at
the pressure of the hydrogasifier. The gas producer for
ore reduction operates on hydrogasifier spent char, with
top and bottom temperatures of 2000 and 3000 F. The ore
reduction and steam decomposition chambers both operate
at 1500 F. This system involves a hydrogasifier, an ox-
idizer of iron where hydrogen is made, a reducer of iron
oxide,   an air-blown gas producer. Consolidation Coal
Co. (Benson, 1970) has suggested the merging of these
four units into two, with hydrogasification and iron oxidation occurring in one, and the solid feed (char and
iron oxide) going to the second unit which combines the
gas producer and iron-oxide reducer. The extent to which
this combination is being considered for replacement of
Hygas has not been established.

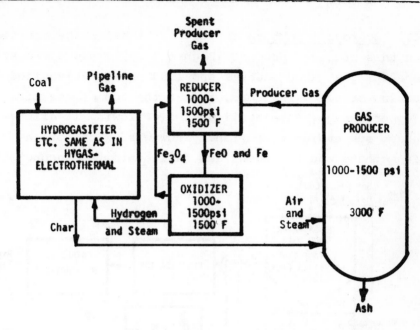

Figure 3-2f. Steam-Iron Process for Making Pipeline Gas from Coal

C3. Hydrogasification (Bureau of Mines). The gasifier
in this proposed process (Figure 3-2g) employs two stages
operating at about 1650 F and 1000 psig. Pulverized
caking or noncaking coal, fed at the top of the first
(upper) stage, devolatilizes while flowing in dilute-
phase suspension concurrently downward with hot gas rich
in methane and hydrogen, and falls into the second stage

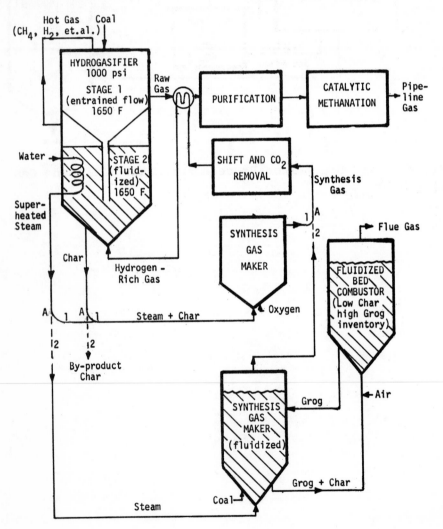

Figure 3-2g. BuMines Hydrogasification Process for Making
Pipeline Gas from Coal

(fluidized-bed) where partial gasification and methanation occur in the presence of almost pure hydrogen. The
high methanation is accompanied by high heat evolution,
and the temperature is controlled by stage-2 coils in
which superheated steam is generated. The hydrogen entering the bottom of stage 2 has alternative sources, both
shown on the flowsheet. At three points marked A (alternatives) flowlines 1 represent the first process (use of
oxygen) and flowlines 2 the second (use of air and solid
carrier). Consider the oxygen method first. Char from
the bottom of stage 2 and superheated steam enter the
synthesis gas chamber along with oxygen, and the products
are subjected to water-gas shift and carbon dioxide absorption, then to heat exchange on the leaving gasifier
gas before entering the bottom of stage 2. In the second
alternative, char from the bottom of stage 2 is treated
as a by-product, and raw coal is used to make hydrogen.
The coal enters the synthesis-gas-making chamber where it
mixes with sufficient hot grog to eliminate trouble from
caking, and reacts with steam. The energy for the endothermic process is supplied by circulation of a grog-char
mix between the synthesis-gas maker and an air-blown
fluid bed which raises the grog temperature by char combustion. The char inventory in the air-blown bed is of
the order of 5%.

The gas leaving the top stage of the gasifier is subjected to the usual sequence of purification and final
methanation.

D1. CSG Process or CO_2-Acceptor (Consolidation Coal Co.).
In this process (Figure 3-2h) lignite (<1/8 to 1/4") is
devolatilized at 140 psia in the presence of steam, carbon monoxide, hydrogen and dolomitic calcine (MgO·CaO)
in a fluidized-bed devolatilizer kept at 1500 F by addition of calcine at 1870 F; char from the latter is fed to
a gasifier bed containing calcine and operating at 1520 F

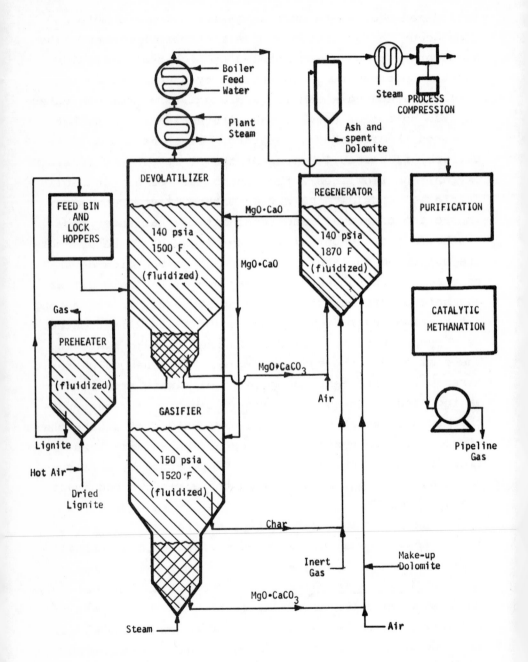

Figure 3-2h. CO_2-Acceptor Process for Making Pipeline Gas from Coal

and 150 psia in a fluidized-bed regenerator which receives separate streams of partially carbonated calcine
($MgO \cdot CaCO_3$) from the devolatilizer and gasifier, returns
regenerated calcine to the same units, and sends waste
gas to an energy recovery system. The circulating solid
material, introduced as dolomite ($MgCO_3 \cdot CaCO_3$), evolves
carbon dioxide with absorption of sensible and chemical
energy in the regenerator, and accepts carbon dioxide and
releases both sensible and chemical energy in the devolatilizer and gasifier. Gas from the gasifier, rich in
hydrogen and carbon monoxide, is fed with steam to the
devolatilizer the gas from which is purified, catalytically methanated, and then compressed.

The process is also designed to operate at about 300
psia, in which case temperatures in the regenerator and
gasifier change to 1940 F and 1575 F and the gasifier operates with a recycle stream.

3.1.4. Stage of Development

Hygas. The Hygas process, supported jointly by the
Office of Coal Research and the American Gas Association,
is the most advanced of the gasification processes. An
80 tons per day (1.5 million cubic feet/day) $8 million
pilot plant located in Chicago is to begin operation in
the summer of 1971, using a Montana lignite and Illinois
high-volatile bituminous coal. The initial operation
will employ hydrogen made by steam-reforming of natural
gas, and the electrothermal reactor for supplying synthesis gas is to begin operation later in 1971.

Two other projects funded by AGA relate to the commer-
cialization of Hygas. The preliminary engineering design
of an 80 million cubic feet/day demonstration plant,
about one-sixth to one-third the size of a commercial
plant, is underway by IGT and Procon Incorporated, and a

study to identify the potential sites in the United
States for commercial gasification plants has been
completed.

Molten Carbonate. The Kellogg molten salt process was
supported by OCR through bench-scale development, but
funding has ceased because no satisfactory material that
is both sufficiently corrosion resistant and economically
feasible has been found to contain the molten salt mixture. This problem has precluded further development
work, although the process has several desirable
features.

Bigas. Research on the Bigas process, which is supported
by the Office of Coal Research, has proceeded through the
operation of a 100 lbs/hr laboratory version of the devolatilization and methanation stage of the gasifier,
with the hot gas from the slagging stage simulated by
combustion of hydrocarbons in oxygen. Coals ranging in
rank from lignite to high-volatile "A" bituminous have
been studied, all without pretreatment, and data for further scale-up have been obtained. The slagging stage of
the gasifier will be simulated for mechanical characteristics and tested for operation at the pilot plant. A
fully integrated 120 tons/day pilot plant is being designed for construction near Homer City, Pennsylvania,
but the construction schedule is uncertain. Federal regulations now require industrial support, which is yet to
be secured, amounting to one-third of the funds required
for the plant before OCR can supply the other two-thirds.
An evaluation of Bigas by Air Products and Chemicals,
Inc., indicates that this process can be competitive with
other gasification processes, and predicts that savings
can be expected from further research.

Synthane. The Bureau of Mines has studied a 4-in. diameter 6-ft long laboratory version of the Synthane gasifier using caking bituminous coals from the Pittsburgh

and Illinois No. 6 seams. The process has received favorable evaluation by the M. W. Kellogg Co. The Bureau of
Mines has arranged for Hydrocarbon Research, Inc., to
study this process using an existing reactor modified to
simulate an 18-in. diameter Synthane gasifier. This gasifier will include separated dilute- and dense-phase
zones consistent with the anticipated commercial design.
The Bureau is planning a $6-7.5 million Synthane pilot
plant, for which a 1000 psi gasifier is now being designed of 4-ft internal diameter and 72 tons coal/day
capacity (1.2 x 10^6 cu ft/day).

Hygas-Oxygen. Several versions of the char gasifier,
usually referred to as the Texaco Process and used for
hydrogen production in this process, have reached various
stages of development. The Texaco Process was in commercial use at 400-500 psi for synthesis gas production at a
rate of 6 x 10^6 cu ft/day (von Fredersdorff and Elliott,
1963). The other parts of the process are essentially
the same as those in the Hygas Electrothermal process of
IGT. That group is committed to the completion of study
of the pilot-scale electrothermal unit being installed in
the summer of 1971, but contemplates possible shift to
oxygen later, and is interested in fluidized-bed operation

Steam-Iron. The continuous steam-iron method for supplying hydrogen to the hydrogasifier is under development at
IGT by Fuel Gas Associates. Fixed-bed steam-iron hydrogen production has been used commercially, and IGT has
developed design information for continuous operation at
pilot-plant scale. The other parts of the overall process are essentially the same as those in Hygas-Electrothermal.

Bureau of Mines Hydrogasification. Research on the hydrogasification process at the Bureau of Mines is still at a
small laboratory stage. Experiments have been performed
with a 3-in. diameter reactor using caking coal from the

Pittsburgh seam, and the results are encouraging enough
to justify further work. This process is at a relatively
early stage of development compared with the other
processes.

CSG (CO_2-Acceptor). The CSG process, funded initially by
Consolidation Coal Co. and now by the Office of Coal Research, will be studied next in a 30 tons/day $9 million
pilot plant being constructed in Rapid City, South Dakota.
The plant is to be finished in the fall of 1971 and gasification runs are to begin in the spring of 1972. Several Northwestern lignites and Western sub-bituminous coals
will be processed, and a number of dolomites and lime-
stones will be tested for use as CO_2 acceptors. The pilot plant simulates only the gasification part of a complete commercial plant. After that part of the process
has been demonstrated, the gas purification and methanation stages can be added using knowledge generated in
other studies. The 30-month pilot plant program will
provide data for commercial design.

3.1.5. Economic Considerations

Economic projections of Synthane and Bureau of Mines Hydrogasification using coal prices corresponding to mine
ownership by the gas plant and equivalent to coal at
$3.31/ton have been given recently by Wellman (1971) and
Mills (1970), respectively. Economics of the other processes here considered were given about two years ago by
Tsaros and Joyce (1968) on the basis of coal at $16¢/10^6$
Btu and lignite at $11.3¢/10^6$ Btu. If the cost estimates
in the latter projects are increased by 10% to account
roughly for today's higher interest rates and construction costs, the total investments for the different
processes at a capacity of 250×10^6 cu ft/day are in
the range $90 to 170×10^6. If the coal and lignite
costs used in both the above studies are adjusted to a

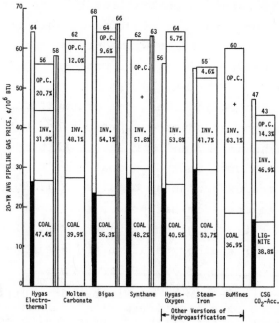

Figure 3-3. Components of Pipeline Gas-from-Coal Price for
Different Processes (OP.C.=Operating Cost; INV.=Investment;
triple vertical line bar=price without sulfur credit; left
narrow bar with coal cost darkened=1971 reanalysis by
Tsaros)

common, updated basis of 17.7¢/10^6 Btu for coal and 12.4¢/10^6 Btu, for lignite, the average 20-year gas prices for coal-based processes range from 55 to 64¢/10^6 Btu, and the percentages of the gas price represented by operating cost, investment, and fossil fuel are distributed as shown in Figure 3-3.* The narrow 3-line bars to the right of the main bars represent the gas price without sulfur credit, which may be more realistic. Very recent estimates by Tsaros (1971) for all the processes except Molten Carbonate and BuMines Hydrogasification give total investments ranging from $106 to $188 x 10^6 and gas prices, for processes using coal, ranging from 54¢ to 68¢/10^6 Btu in 1970.** These revised prices are shown as narrow bars (white above black) at the left of each main bar. Although they purport to be based on coal at 16¢/10^6 Btu and lignite at 11.4¢/10^6 Btu, the contribution of fuel to the total cost is in fact consistent with our own adjustment of fuel costs to 17.7¢/10^6 Btu for coal and 12.4¢/10^6 Btu for lignite.

Investment here includes coal preparation and gasification, oxygen and hydrogen production, gas purification and methanation, and offsite facilities. (Some accounting procedures include a coal mine as part of the investment and oxygen as an offsite facility.) An investment breakdown for six processes, reported by Tsaros and Joyce (1968), is given in Figure 3-4. Investment proportions for coal preparation and gasification, plus

----

*The coal cost for Bureau of Mines Hydrogasification is here estimated as a fraction of the Synthane coal cost, which fraction is the ratio of their coal feeds.

**This range of prices would change to 77-97¢ by 1976 if an overall escalation of 6% per year is applied. Subsequent to completion of this report two new estimates have become available for pipeline gas from coal in 1976--83¢ to $1.01 and 88¢ to $1.15.

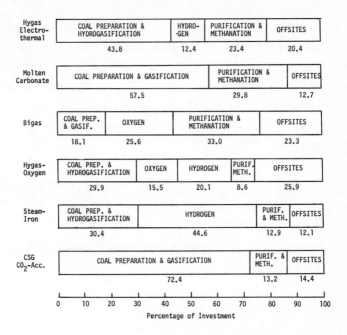

Figure 3-4. Comparative Investment Breakdown for Different Pipeline Gas Processes (Tsaros and Joyce, 1968)

costs for oxygen and hydrogen, range from 43.7% for Bigas to 72.4% for CSG, while the proportions for purification and methanation range from 8.6% for the Texaco version of hydrogasification to 33.0% for Bigas.

The sensitivity of the gas price to the costs of coal or lignite would be measured by the reciprocal of the plant thermal efficiency but for the differing by-products of the different processes. A thermal efficiency of 2/3 calls for a 1.5¢ increase in gas cost for every cent increase in coal cost, both on a Btu basis. The bar graphs indicate that equipment and fossil fuel costs are major contributors; operating cost is minor. Labor varies between 2 and $3.5¢/10^6$ Btu gas, reflecting the high degree of automation expected in pipeline gas plants. An increase of about $10¢/10^6$ Btu in the gas price results from an increase of $1.5/ton in coal cost or an increase of 40×10^6 in plant cost.

Several of the processes show substantial by-product

credit for fuel, power, and sulfur. The percentages of
the gas price attributed to the three cost components
shown in Figure 3-3 are based on by-product credit applied to the plant costs as a whole. Reductions in gas
price (¢/10^6 Btu) due to by-products are summarized as
follows (Nos. 1 to 4--Tsaros and Joyce, 1968; No. 5--
Wellman, 1971):

1.

Hygas-Electrothermal and -Oxygen: 6.3¢ from by-product
fuels and 2¢ from sulfur at $20/ton.

2.

Molten Carbonate: 8.3¢ from by-product power at 5 mills/
kWh.

3.

Bigas: 2.1¢ from sulfur at $20/ton.

4.

Steam-Iron: 3.8¢ from by-product fuels.

5.

Synthane: 9.7¢ from by-product char and tar and 1.2¢
from sulfur.

The value of estimates of the cost of making pipeline quality gas from coal may be questioned on a
number of grounds: During the development stage the
various processes are competitive, and there is a tendency to play the game of not arriving at a cost figure
which is significantly above that of competing processes; an estimated price too far above current gas prices
would tend to dry up federal support of development;
each process is associated with a different contractor
in the design of its plant, with correspondingly different techniques used to estimate plant costs and,
particularly, contingencies; the dates of the different
designs are different, and in a period of monetary inflation this affects the comparison. Considering the
narrow range of variation in estimated gas cost, the

unreliability of the projections for the above enumerated reasons, and the fact that up to now no data have been obtained from pilot plants of any one of the processes, it is clear that no good economic base exists for present choice among them. Considering the hazards of projections with inadequate data, it is probable that, regardless of process, pipeline-quality gas from coal in 1976 will cost, at its point of production, in excess of 85¢ and more probably in excess of $1, at least until a commercial process optimization has occurred.

3.1.6. Comparison of Processes

Although ranking the different proposed schemes on economic or technical grounds would be premature, comparison of the performance, advantages, disadvantages, and questionable aspects of each scheme serves as one type of comparison and, at the same time, provides a basis for assessing research needs. Such comparative information is summarized in Table 3-1 using methane yield as one of the measures of performance. To aid in the comparison, a montage of Figures 3-2 a-h is reproduced following Table 3-1.

The table does not include a comparison of temperatures used in the devolatilization and noncatalytic methanation sections of the various processes. Synthane uses the lowest devolatilizing temperature of 1100-1300 F, the Hygas schemes (electrothermal and oxygen) and Steam-Iron 1300-1500 F, CO_2 Acceptor 1500 F, Bigas 1400-1700 F, Bureau of Mines Hydrogasification 1650 F, and Molten Carbonate 1830 F. The lowest temperature consistent with adequate rate is desirable, and the processes are not sufficiently similar to identify quantitatively the trade-off of rate versus thermodynamic tendency to make methane. Perhaps this is one more way of saying the processes cannot be evaluated

Table 3-1. Comparison of Processes for Making Pipeline Gas from Coal (Footnotes at end of Table)

ADVANTAGES	DISADVANTAGES	QUESTIONABLE AREAS
HYGAS-ELECTROTHERMAL [Coal Size = -1/8 inch; Pressure = 1000 to 1500 psi; Methane Yield on Three Bases *: A = 0.33; B = 1.69; C = 0.83]		
Highest noncatalytic methanation (i.e., in gasifier) of any process developed beyond bench-scale.	The pretreating operation necessary to handle caking coal produces an extra gas stream and prevents making full use of the relatively high reactivity of fresh coal.	Electrothermal gasification** is economically questionable (the electric energy need is of order of one-sixth the energy in the final pipeline gas). Electrode life may be short in large systems due to erratic bubble action at electrodes.
Slurrying of fuel gives the reliable feed to a high-pressure system that characterizes liquid pumping.	Disposition of by-product char adds a second stream to be disposed of and a degree of uncertainty about economics of disposal.	Fluid-bed scaling problems may arise, with less reaction achieved in large diameter bed.
Hydrogasification appears, from its effect on methane yield (theoretical or experimental), to be superior to energy liberation by oxygen directly in gasifier.		Pretreater temperature control without runaway or localized agglomeration may present difficulties, possibly related to distribution of pretreater air input.
MOLTEN CARBONATE [Coal Size = -12 mesh; Pressure = 420 psia; Methane Yield on Three Bases *: A = 0.098; B = 0.50; C = 0.29]		
Uses air instead of oxygen, and minimizes mixing of flue gas with synthesis gas.	Molten salt is very corrosive. Satisfactory containment has not been achieved.	Control of melt circulation within and between vessels may be difficult.
Physical properties of feed, particularly size, not critical because of suspension in salt melt; coal preparation costs less.	Temperature of gasifier too high for significant methanation to occur, and so high that methane from coal devolatilization is partly decomposed. This behavior is evidenced by the process giving the lowest yield of methane from the gasifier on any of the three bases A, B and C given above.	Entrainment of melt and disengagement of gases from melt may be difficult to retard and promote, respectively, in large scale equipment.
Significant part of sulfur is removed by carbonate and collected in ash removal process.	Conversion of bicarbonate to carbonate in gasifier evolves carbon dioxide and consumes energy.	Further development awaits solution of the salt corrosion problem.
The sodium carbonate is claimed to have advantageous catalytic effect on rate of solids gasification (19-35 lbs/ft³ hr at 3 atm, for sizes 0.16mm to 3.6 mm, vs 35 lbs/ft³ hr for Synthane at 40 atm).		

Table 3-1. (Continued)

BIGAS

[Coal Size = -200 mesh; Pressure = 750 to 1500 psi; Methane Yield on Three Bases*: A=0.21; B=1.04; C=0.52]

ADVANTAGES	DISADVANTAGES	QUESTIONABLE AREAS
Presumably a new concept as of 1965, and a result of a state-of-the-art survey.	Reactor for steam-oxygen-char is much hotter than in other processes, with attendant higher thermal loss in slag (about 1% of energy in fuel fed).	Fuel inventory in the top gasifier, affecting residence time, is unknown and dependent on flow pattern and solids recycle ratio, also unknown.
Entrained gasification is fast and simple if residence time is adequate.		Adequacy of high-temperature high-pressure cyclone separator is not established; there is possibility of difficulty with nearly burned-out low-density particles.
Gasifier methanation, as fraction of entering fuel-carbon or of methane-equivalent of entering fuel-hydrogen, is claimed to be higher than other oxygen-blown processes.		Slag-tap control at high pressure may be troublesome.
		Ability to dodge lock-hopper feed by use of piston displacement and that solids-flow constancy which is demanded by the small system-holdup are not established.
		Process hinges on assumption that devolatilization and some methanation can occur fast enough to eliminate need for fluidized bed. Residence-time distribution is important, unknown and probably very different from bench to pilot to demonstration scale.
		Complete throw-out of slag in hot stage 1 is necessary to prevent agglomerate buildup on walls of cooler stage 2.

SYNTHANE

[Coal Size = 70% thru 200 mesh; Pressure = 600-1000 psi; Methane Yield on Three Bases*: A=0.18; B=0.79; C=0.55]

ADVANTAGES	DISADVANTAGES	QUESTIONABLE AREAS
Direct use of caking coal originally claimed feasible by free fall of coal with steam-oxygen in top chamber, thereby eliminating cost of a separate pretreater and loss of gas.	Production of char stream as one of the final products requires establishing a market or operating a power plant. Char may be difficult to burn.	It now appears that the analogue of pretreating, the free fall in a top section above the gasifier, must be modified to a fluidized bed operation to permit coarser grinding and higher throughput. Operability of this concept has not been established.
Parallel flow at feed point tends to minimize loss, by cracking, of evolved hydrocarbon.	Amount of catalytic methanation necessary is much higher than in hydrogasification processes.	The scale-up factor in a dense bed is not established, particularly the effect of large diameter in fluidization and the provision for transition between stacked beds of different density.
		Gasification rate per unit area of bed is 200 lbs coal/ft²hr in laboratory experiment; 400 is hoped for in commercial plant.
		There may be a problem of distribution of oxygen feed in large-scale unit to prevent localized high temperature and associated agglomerate formation.

Table 3-1. (Continued)

ADVANTAGES	DISADVANTAGES	QUESTIONABLE AREAS
	SYNTHANE (Continued)	Pretreater temperature control (see Hygas-Electrothermal).
		Problem of how to mount dilute-phase bed below dense-phase bed of larger cross section, without segregation or distribution problems, is unsolved.
		Steam conversion in laboratory experiments is about 30%; 50% is desired.
		Carry-over of solid carbon may be high with conventional pulverized coal sizes.

[Coal Size = -1/8 inch; Pressure=1000 to 1500 psi; Methane Yield on Three Bases*: no data]

HYGAS-OXYGEN

ADVANTAGES	DISADVANTAGES	QUESTIONABLE AREAS
Over Hygas: Oxygen replaces electrical energy, and there is possibility of adjusting the hydrogen/carbon monoxide ratio to a more favorable one for methanation by removing more oxygen in absorbed carbon dioxide than was put in as oxygen.	Direct coupling of the hydrogen-carbon monoxide supply to the methanation process is absent, with attendant thermal loss.	Merits of electrogasification vs use of oxygen, water-gas shift, and purification are not settled. Heating of hydrogen-rich gas fed to gasifier may be required.

[Coal Size = -1/8 inch; Pressure=1000 to 1500 psi; Methane Yield on Three Bases*: A=0.25; B=1.27; C=0.64]

STEAM-IRON

ADVANTAGES	DISADVANTAGES	QUESTIONABLE AREAS
Over Hygas: Air replaces electrical energy, and there is possibility of substantial reduction in equipment cost.	Direct coupling of the hydrogen supply to the methanation process is absent, with attendant thermal loss. Provision of hydrogen at the expense of discarding carbon monoxide and some hydrogen favors formation of carbon monoxide and hydrogen over carbon dioxide and methane, thereby resulting in lower methane yield in the gasifier than is obtained in Hygas or Bureau of Mines Hydrogasification. Requires sulfur removal from reductor effluent.	Continuous, high-pressure, large-scale hydrogen production by the steam-iron technique is not well established. For example, are there problems of dusting, agglomeration in the producer, agglomeration in the ore bed, and loss of ore activity upon cycling. Existence of six fluidized beds (3 in gasifier, 3 in steam-iron) in balanced operation presents serious control problems.

Table 3-1. (Continued)

ADVANTAGES	DISADVANTAGES	QUESTIONABLE AREAS
BUREAU OF MINES HYDROGASIFICATION		
[Coal Size = 50x100 mesh (in lab studies); Pressure=1000 psi; Methane Yield on Three Bases*: A=0.50; B=2.90; C=0.95]		
Air is used instead of oxygen in energy production. Caking coals can be used without pretreatment. High conversion of coal volatiles to methane and substantial methanation within gasifier; therefore relatively small amount of catalytic methanation is required.	Requires sulfur removal from flue gas leaving fluidized-bed combustor.	Cooling the fluidized-bed section may present a scale-up problem.
	Absence of direct coupling of the hydrogen-producing process to the methanation process introduces a thermal loss which may offset the methanation gain.	Operation of fluidized bed containing grog and circulation of grog are not yet developed.
Offers method for decoupling hydrogen supply from methanation process and for elimination of steam-carbon reaction in chamber devoted to methanation only.	Well behind other processes in its stage of development.	Carry-over of solid carbon may be high unless small-size fraction is not used.
		How to arrange the carbonization chamber above the fluidized bed and how to inject gas from the fluidized bed into the carbonization chamber have not been established. There is little present experimental support of the process design.
CSG (CO₂ ACCEPTOR)		
[Lignite Size = 1/4 to 1/8 inch × 0; Pressure=140 psia; Methane Yield on Three Bases*: A=0.16; B=0.90; C=0.46]		
Carbon dioxide removal in gasifier favors methanation.	Tendency of acceptor plus steam to form a melt at high temperature and steam pressure prevents T from exceeding 1500-1550F. Lignite and some subbituminous coals are the only ones active enough to meet this temperature restriction.	Sulfur dioxide removal from regeneration effluent is probably necessary,but not yet included in flow sheets. (This gas stream contains 2300 ppm sulfur dioxide when the lignite contains 0.59% sulfur.)
Uses air instead of oxygen.		
Supplies reaction energy in situ.	Regenerator outlet gas must contain 3-4% CO to prevent CaS formation and sticking.	System control may be troublesome;bed densities must be maintained to control stone showering time through beds; char inventory in regenerator must be kept at about 5%; gasifier temperature must be kept below incipient stone-calcine melting; local hot spots in regenerator could produce agglomerates; energy balance between gasifier and regenerator could go off due to depth, density, or char fraction in either bed.
Given lignite as fuel, this process appears to offer minimum cost.	Noncatalytic methane yield is relatively low.	

*   All based on feed of Illinois No. 6 coal, except CSG (Renner Cove Lignite), Molten Carbonate (Pittsburgh seam coal) and Bureau of Mines Hydrogasification (Pittsburgh seam coal). A=(Methane leaving gasifier)/(Carbon in solids feed stream to gasifier); B=(Methane leaving gasifier)/(Methane-equivalent of hydrogen in coal); C=(Methane leaving gasifier)/(Methane in final pipeline gas). Yields based on mass flows given by Henry and Louks (1970), except for Bureau of Mines hydrogasification which is based on data given by Mills (1970).

** I.G.T. is prepared to replace electrothermal gasification with another method of hydrogen production if necessary.

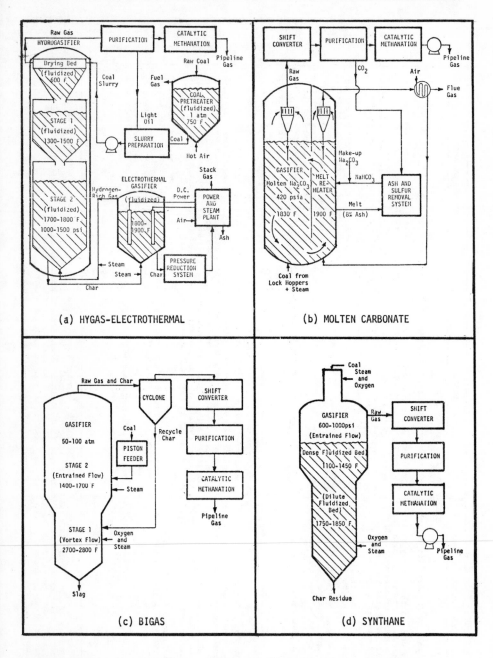

(a) HYGAS-ELECTROTHERMAL

(b) MOLTEN CARBONATE

(c) BIGAS

(d) SYNTHANE

Figure 3-2. Processes for Making Pipeline Gas from
Coal [(a) Hygas-Electrothermal; (b) Molten-Carbonate;
(c) Bigas; (d) Synthane; (e) Hygas-Oxygen; (f) Steam-
Iron; (g) BuMines Hydrogasification; (h) CO_2-Acceptor]

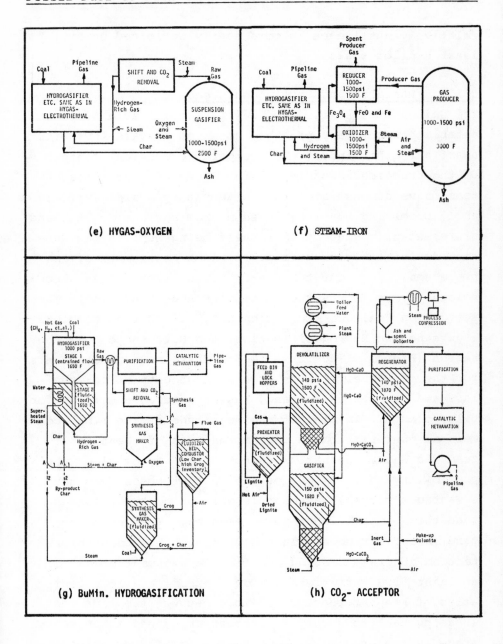

(e) HYGAS-OXYGEN

(f) STEAM-IRON

(g) BuMin. HYDROGASIFICATION

(h) CO$_2$- ACCEPTOR

relatively until each is more nearly optimized on at
least a pilot scale.

3.1.7. Research and Development Needs

The United States will need gas from coal, at a time not
established but very probably soon, on a scale that will
dwarf all other industry of comparable chemical content.
Of the processes described here, those of four research
teams have demonstrated--on a laboratory scale--suffici-
ently promising results, freedom from serious faults, and
differences in concept to warrant being kept in the running during the period of pilot-plant assessment of the
processes. The four teams are the Institute of Gas Technology (Hygas in its two variations and Steam-Iron), Consolidation Coal Co. (CO_2 Acceptor), U.S. Bureau of Mines
(Synthane and Hydrogasification) and Bituminous Coal Research (Bigas). The first two of these have pilot plants
for the main gasification process near completion; the
last two have plans for and, research-wise, are ready for
pilot plants. In view of the magnitude of the final commercial operation based on the research of these four
teams to date, the nation cannot afford to overlook the
possibility that any one of the four processes may have
advantages properly assessable only on a larger scale
than that of past laboratory research. The analogous
case in nuclear power-plant development has been the retention at great cost during the development stages of
at least five varieties of slow reactors and four varieties of breeder reactors, simply because the time lost
in going back to a complex process abandoned in an early
stage is too great. Viewed another way, a retention of
four processes through sufficiently advanced stages to
permit making a final choice of process that saves 10¢/
1000 cu ft in manufacturing cost will, in the days when

our consumption of synthetic pipeline gas is one-third
that of natural gas in 1970, amount to a saving of $750
million per year. We cannot afford to decide prematurely
on the wrong process.

The strong recommendation of continuation of four processes at least through the pilot-plant stage implies
that much basic research has already been carried out on
coal gasification, but it does not imply absence of the
need for more. Any process which wins the race will be
in need of continuous modification and improvement; and a
strong research program, both basic and applied, on the
physics and chemistry of coal gasification is needed.
The proposed research is divided into gasification fundamentals, physical behavior of coal and ash during gasification, entrained flow, fluidized beds, gas purification
and methanation, solids feeding, and hydrogen production.
Gasification Fundamentals. The evolution of gas from a
heated coal particle is a complex problem of interaction
of heat transfer, mass transfer, and chemical kinetics
not really understood despite numerous papers in the
area. The effects of temperature, particle size, ambient gas concentration, and total pressure on (i) the
kinetics of volatile evolution, char gasification, and
cracking reactions exhibited by primary volatile products and (ii) the factors controlling the final gas
characteristics such as tar content should be established on a quantitative basis.

The expected advantages of flash heating finely-
ground coal should be explored more fully, with special
emphasis on possibilities for increasing the extent of
volatilization, methanation, and solids desulfurization.
In order to permit more accurate prediction of equilibrium gas composition, the free energy of formation of
fresh coal char relative to that of graphite and the
dependence of this property on gasification conditions
and coal type should be determined. The study should

also determine the effects of ash content of the coal
and composition of the ash on the char gasification rate
and, in relation to that effort, should identify materials that catalyze the gasification at relatively low
temperatures where equilibrium methane yields are high.

A generalization of the gasification behavior of a
number of different coals, lignites and by-products chars
from other synthetic fuel processes should be sought.
The observed influence of coal type on gasification behavior should be generalized to the extent possible,
using as a measure of coal type its proximate analysis,
ultimate analysis, petrographic composition, or some
other properly yet to be identified. With regard to
petrographic composition, research on the gasification
behavior of individual petrographic constituents is
strongly recommended.

Physical Behavior of Coal and Ash in Gasifiers. The tendency of coal particles to form cenospheres during heating is known to depend upon heating rate, final temperature attained, ambient gas composition, and coal type;
particle size and total pressure are probably also important. At a certain stage the particles agglomerate
to a degree which depends upon the operating conditions.
During gasification, ash particles are formed with
sizes ranging from that of the individual elements of
inorganic material in the original coal to agglomerates
including ash from several coal particles. The above
phenomena bear significantly upon gasification behavior
and on design features for removing ash and avoiding
the need for pretreatment, but understanding is presently at best only qualitative. Research on single
particles and clouds of particles should be conducted
to determine the history of coal particles and ash
constituents during the gasification and ash-removal
steps. These studies should include both dry-ash and
slagging conditions.

Entrained Flow. The advantages of entrained flow in
gasification processes are well established, but research is needed on some problems inherent in this technique. The operation of entrained-flow systems often
involves mixing a stream carrying raw coal or partially
gasified char and volatile products with hot gas from a
char gasification zone, then separating the gas product
from particles of residual char. The mixing must be
rapid, with minimum back mixing; this requires pro-
motion of turbulence while controlling the major flow
pattern. The separation step may be difficult because
of the relatively small density difference between high-
pressure gas and char particles in the form of cenospheres. Research on these mixing and separation operations would be highly valuable in the development of
equipment.

When chemical and/or physical reactions are carried
out in a suspended-solid-gas system in vortex flow,
knowledge of the distribution of residence times as
affected by mode of introduction of the feed streams,
by axial momentum flux and angular momentum flux of the
feed, by particle size and size distribution, and by
density of the gas and particulate matter is of high
importance, especially when the reaction is fast. A
thorough generalized study of this flow and transport
process is warranted.

Fluidized Beds. It is evident from the proposed schemes
for making pipeline gas from coal that considerable use
of fluidized-bed gasifiers is expected. The behavior
of fluidized beds with regard to bubble formation,
mixing, and heat transfer is highly important in determining gasification behavior but is not well understood
for high-pressure operation. Since this behavior,
particularly local segregation or local surging, is
probably very dependent on bed diameter, model research
on this problem has its difficulties and large-scale

studies may be necessary. The ability of high-pressure
fluidized beds to use caking coals has not been established. Possible techniques being studied for overcoming the problems caused by caking coals include pretreating the coal prior to addition to the bed (Lee,
1970) and using an inert power (grog) mixed with the
coal in the bed (Mills, 1970). Pretreatment has the
disadvantages of not making full use of the relatively
high reactivity of the fresh coal and of producing an
extra gas stream to be handled; the grog technique introduces an extra solid-handling problem. More research
on handling caking coals within the gasifier is needed.
The interesting proposition of conducting only methanation (exothermic) in the fluidized-bed gasifier requires a bed-cooling technique that is not well identified. Small-scale research in these areas would be
valuable in determining general principles, but the behavior in question is undoubtedly strongly sensitive to
scale, and research on large systems may be necessary.
Gas Purification and Methanation. The product gases
from the different types of gasifiers require essentially the same type but different extents of treatment.
The purification steps are relatively well established,
but there is some uncertainty about the behavior of
processes for removal of sulfur compounds and carbon
dioxide at the high pressures of interest. Since these
operations contribute a significant fraction of the
overall cost of gas production, efforts to improve
available techniques are justified.

Some processes require sulfur removal from gas
streams other than the product gas. Sulfur in the coal
pretreatment effluent must be reduced to an acceptable
level either before or after the gases are burned, and
the flue gases from spent char combustion will require
sulfur dioxide removal. Elimination of the need for
pretreatment and simultaneous enhancement of sulfur

evolution during volatilization and noncatalytic methanation in processes employing spent-char combustion
would allow a larger fraction of the sulfur to be removed in one operation. Research along these lines
should be the main approach to this problem, since research on stack-gas cleaning is pursued in another area.

Catalytic methanation is also an important cost item.
This step is not well established, and requires considerably more research and development. The reaction con-
verts carbon monoxide and hydrogen to methane and water
vapor, liberating about 65 Btu per SCF of feed gas converted. The importance of efficient temperature control is emphasized by the unusually high adiabatic temperature rise associated with completion of the re-
action--4000 F even before allowance is made for the
reactant preheating that is associated with counterdif-
fusion through the boundary layer over the catalyst. A
quite sophisticated model is warranted to minimize the
chance of catalyst deterioration by sintering and carbon
deposition.

Several types of catalytic methanation reactors have
been studied:

a.

Fixed-bed, cooled by heat-exchange surfaces (Dent and
Hebden, 1949; Dirksen and Linden, 1963);

b.

Fixed-bed, cooled by gas recycle (Dent et al., 1945;
Forney et al., 1965; Wainwright et al., 1954);

c.

Fluidized-bed, cooled indirectly by heat-exchange surfaces (Dirksen and Linden, 1960; Schlesinger et al.,
1956); and

d.

Tube-wall, with catalyst supported on and cooled by
heat-exchanger tubes (Gilkeson et al., 1953; Field and
Forney, 1966).

The most effective catalyst seems to be nickel, but
iron also works. The Bureau of Mines (Haynes et al.,
1970) has studied Raney nickel catalyst (42% Ni - 58%
Al, followed by leeching out 85% of the Al) and IGT
(Lee and Feldkirchner, 1970) has studied commercially
available nickel on Kieselguhr and nickel on alumina.
The two groups achieved catalyst lifetimes in excess of
2500 and 1000 hrs respectively, in small equipment, but
there is some indication that it may be more difficult
to achieve acceptable lifetimes in larger equipment
(Hayes et al., 1970).

Some principles and constraints governing the design
of catalytic methanators are the following:

a.

Sulfur poisons nickel catalyst; it is usually assumed
that the H_2S content of the feed gas must be below about
0.1 ppm. Nickel carbide is also a poison, and arsenic
may cause difficulty, but its expected concentration
and effects are not well understood.

b.

The most desirable operating temperature is about 750 F.
Temperatures as high as possible, but definitely above
550 F, are desirable in order to allow the heat gener -
ated to be used in making superheated steam. Nickel
carbide forms at temperatures below about 665 F; the
time for heatup must be minimized, or heatup must be
conducted in the absence of CO. Temperatures above
about 850 or 900 F result in carbon deposition on the
catalyst, and catalyst sintering occurs at temperatures
not much above 900 F.

c.

Recycling with water removal improves equilibrium
methane yields.

d.

The final carbon monoxide composition must be less than
1000 ppm.

Laboratory research on catalytic methanation should be designed to achieve better definition of the necessary operating conditions. More bench-scale studies are needed for developing better pilot-plant equipment, and the best designs should be tested at local full scale and on real rather than simulated pilot-plant gas.

Solids Feeding. A problem common to all the proposed processes is feeding solids into a high-pressure vessel. The required feeder must be large-scale and capable of steady, controllable operation. Techniques under study include lock hoppers, slurry pumping, and piston displacement. Some versions of the first two techniques are known to be troublesome under certain conditions and they may require more development to be successful in gasification plants. The third technique is relatively new and its performance is yet to be established. Since solids feeding is a crucial part of the gasification system, development of a suitable feeder is clearly important.

Hydrogen Production. Several alternatives are available for using process char to make the hydrogen or synthesis gas fed to the hydrogasifier, including the steam-iron process, partial oxidation, and electrothermal gasification. The best alternative for a given gasification scheme depends upon the way CO and CO_2 additions to a hydrogen-steam mix contribute to its reactivity with char and on the economics of joint operation of a power plant and a gas plant. The steam-iron process has been used commercially in fixed-bed, low-pressure operation (Nelson et al., 1966), but continuous, high-pressure operation is yet to be established. Partial oxidation has been used industrially for synthesis-gas production and it should be adaptable to large-scale hydrogen production. The electrothermal technique has been studied in small-scale equipment (Ballain and Pulsifer, 1969; Kavlick et al., 1970), but technical feasibility

has not been established and the economics are questionable. Research is needed to define the principles of
hydrogen-process selection, to identify the most economic methods of hydrogen supply, and to make the viable
alternatives operational under the conditions required
in gasification.

3.2. Minimum-Cost (Low-Btu) Clean Gas from Coal

3.2.1. Background
Gas producers, the technical name for devices in common
use fifty years ago for making low-Btu nitrogen-diluted
gas, have almost disappeared from the American scene.
They were once used primarily in close coupling to the
furnace to which they supplied fuel, thereby allowing
effective delivery of the sensible heat content of the
hot fuel gas; but they were also sometimes used to make
cleaned cold gas. The reasons for disappearance of the
gas producer were:
1.
Its production of dirty gas--high in dust, soot and tar--
which is objectionable in many application;
2.
Difficulties in bed clinkering from poor temperature
control;
3.
Sensitivity to coal type, i.e., occasional need for hand
poking when caking coals were used, even though the fuel
bed was mechanically stirred;
4.
Sulfur content of the gas, objectionable in some applications even fifty years ago;
5.
Restriction to small size, with associated high cost of
labor and equipment for large-Btu use;

6.
The increasing availability of cheap natural gas.

Evidence is clear that the last reason--the one of dominant importance--is on the point of disappearing in the
United States, and the desirability of again being able
to make gas from coal arises.

The impetus for developing low-Btu gas from coal
comes mainly from the electric power industry, but partly from heating and processing operations which once
used gas from coal but now use natural gas. These
operations might profitably switch back to low-Btu gas
as natural gas becomes more costly. With our change in
standards of acceptability of equipment and processes--
automatic character, freedom from hand labor, large
scale of operation, acceptable environmental effects--
it is clear that the first five above-enumerated reasons
for disappearance of gas producers must be removed. If
the gas producer is to come back, it must be very different from its ancestor.

3.2.2 New Design Possibilities

First-Generation Processes. Using technology now available, first-generation processes for making low-Btu gas
from coal can be constructed by employing a conventional
fixed-bed gas producer followed by a gas purification
system. The sulfur content of the gas would have to be
less than about 450 g/10^6 Btu, which corresponds to
about 1000 ppm sulfur dioxide in combustion products resulting from burning the gas with 10% excess air. Since
techniques now available for sulfur removal require that
the gas be cooled and cleaned of dust, soot and tar,
first-generation processes overcome the dirty-gas problem mentioned above, but the need to conserve sensible
heat introduces a heat exchange problem not required

in previous gas producers.

A first-generation process emplying a Lurgi gas
producer (Figure 3-5a) and serving a combined gas turbine-steam turbine power plant is analyzed in a recent
report by United Aircraft Corporation (Robson et al.,
1970). Although the Lurgi producer is known to have
difficulty using American caking coals, the proposed
scheme is outlined below because (a) its gas purification system demonstrates a possible arrangement for conventional producers, (b) the UAC analysis provides an
indication of the process economics, and (c) the Lurgi
design illustrates the principles of fixed-bed gas pro-

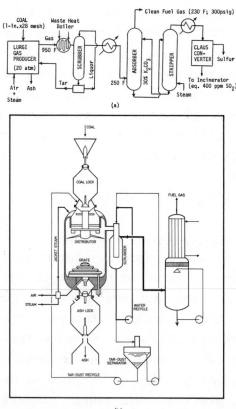

Figure 3-5. Lurgi-Based Process for Making Clean Fuel Gas
from Coal (upper part-Flow Sheet Showing Gas Producer,
Waste Heat Boiler and Gas Purification System; lower part-
Lurgi Pressurized Gas Producer with Hot Tar Scrubber)

ducers. A Lurgi plant of the general type here considered was recently built in West GErmany (Rudolph,
1970a).

Coal is crushed to pass a 1-in. screen and the fraction less than 28-mesh is briquetted with process tar
and fed, with the 1-in. x 28-mesh fraction, through a
lock hopper to the gasifier, shown in Figure 3-5b. The
gasifier is essentially a refactory-lined water-cooled
cylindrical shell operating at 20 atm in which air and
steam introduced at the bottom pass upward through a
7-ft (ca.) deep fuel bed maintained by feeding raw coal
at the top. The feed rates are 0.8 lb steam and 2.7
lbs air per lb coal (MAF). At the bottom of the bed
carbon burns with air to form carbon dioxide in a thin
combustion zone (at 1750-2100 F) in which, simultaneously,
the endothermic reaction of steam with carbon to form
carbon monoxide and hydrogen holds the temperature below the ash fusion temperature, thereby controlling
clinker formation. As the hot combustion gases flow
upward through the bed, carbon monoxide is formed by
the endothermic reaction of carbon dioxide with carbon;
the lightly endothermic water-gas shift reaction
(Equation 2, Section 3.1.1) maintains substantial equilibrium among CO_2, CO, H_2O and H_2; and some hydrogen
reacts exothermically with carbon to form methane.
Fresh coal undergoes successive drying, devolatilization, and reaction with oxygen and steam, and the vola-
tiles crack to form hydrogen, methane and higher hydrocarbons. The net reaction occurring in the thick part
of the bed above the combustion zone is endothermic,
and the gases leave the top of the bed at about 950 F.

The gasification efficiency at the producer outlet
is about 95%, the losses comprising 1-2% loss due to
unburnt carbon in the ash and 3-4% heat losses.

In the UAC study the gases leaving the gasifier are

cooled to 550 F in a waste-heat boiler, raising about
60% of the steam required for gasification. At the
estimated tube temperature of 450 F it is claimed that
the tar condensing on the heat transfer surfaces of the
boiler will be free-flowing and relatively stable. (However, tar runoff--especially in the presence of suspended soot--may be in question.) The use of a hot tar-
scrubbing tower (Lurgi recommended practice, Figure 3-5b)
with heat recovery is possible, but the subject report
concludes that the cost exceeds that of a simple waste-
heat boiler.

Residual tar and dust are next removed from the gas
in a water scrubber where the gas temperature is reduced
to 290 F. Scrubber off-gases are further cooled to
250 F in a water-cooled heat exchanger.

Sulfur compounds (H_2S and COS) may be removed by any
of a numberof wet or dry processes, but the particular
flow sheet here considered employs hot carbonate scrubbing (30% K_2CO_3 at 230 F). Data indicate that 85% of
the H_2S and COS and 12% of the CO_2 are removed. This
reduces the sulfur content of the fuel gas to 220 $g/10^6$
Btu (equivalent to about 550 ppm SO_2 in flue gases),
well below the first-generation goal of 450 $g/10^6$ Btu.
AS much as 99% of the sulfur can be removed at extra
expense in the same type system, modified. The fuel gas
then passes through a high-efficiency entrainment separator, not shown in the diagram, with recycle of the
K_2CO_3 liquor, and is then delivered to the power station
at 230 F and 20 atm.

The spent carbonate liquor is steam-stripped and recycled. The gas effluent from the strippers, containing
approximately 10.9% H_2O, 67.5% CO_2, 21.6% H_2S and minor
inert gases, is then cooled to 140 F and passed to a
Claus converter where 90% of the sulfur is recovered as
elemental sulfur. In this process one third of the H_2S
in the gas is burned to produce SO_2, and the effluent

gas is cooled and passed through two bauxite catalyst
beds, operating in series, where SO_2 reacts with H_2S
to produce sulfur. Sulfur compounds not removed are
incinerated, producing stack gases containing about
400 ppm SO_2.

The report (Robson et al., 1970) estimates the performance and economics of a plant producing fuel for a
nominal 1000MW combined gas turbine-steam turbine power
plant with 70% operating factor. It would employ 23
12-ft-diameter gasifiers,* each with its own waste-heat
boiler and water scrubber, and a battery of 6 12-ft-dia-
meter absorption columns operating in parallel, each
preceded by its own heat exchanger. The plant would
consume approximately 6670 tons/day of coal and produce 703 x 10^6 SCF/day of clean fuel gas and 160 tons/
day of by-product sulfur. The energy content of the
gas** is about 77% of that of the coal used. The cost
of producing clean fuel gas is here given on three different bases: cost per kWh; cost per 10^6 Btu of clean
fuel gas; and cost per 10^6 Btu of coal. These three
are related by the efficiencies of different parts of
the overall process:

$$\text{Efficiency of gas plant} = \frac{\text{Btu in clean fuel gas}}{\text{Btu in coal}} = 0.77,$$

$$\text{Efficiency of electrical plant} = \frac{\text{net kWh}}{\text{Btu in clean fuel gas}}$$

----

*Lurgi contemplates a diameter increase from 12 to 16 ft,
and claims an associated probable gasifier capital cost
reduction, per unit capacity, of 25% (Rudolph, 1970b).

** Higher and lower heating values are 173 and 156
Btu/SCF; sensible heat above 60 F is 3.3 Btu/SCF.

Efficiency of electrical plant = 0.45*

Overall efficiency = $\dfrac{\text{net kWh}}{\text{Btu in coal}}$ = 0.35.

The total capital cost for the gas-making process is about $45 x 10^6 or $48/net kW (net power is 918 MW). (If the clean fuel gas were burned in a conventional steam plant having 38% electrical efficiency (and therefore 740 MW in size), the gas-making capital cost would be $57/kW.) Capital costs are distributed roughly as follows:

Gasification	45%
Air Compression	20
Coal handling and preparation, mainly briquetting	11
Acid-gas scrubbing and sulfur recover	10
General Facilities	9
Additional items	5
	100%

The total annual operating plus capital cost attributable to the gas plant plus the coal, with coal valued at 20¢/10^6 Btu ($4.40/ton as received) and a sulfur credit of $25/long ton, is 4.15 mills/kWh. This cost corresponds to 42.6¢/10^6 Btu coal or 54.7¢/10^6 Btu gas. (If the gas were burned in a conventional plant at 38% electrical efficiency, the total annual capital plus operating cost would then be 4.91 mills/kWh.)

Eliminating the sulfur credit of 0.23 mill/kWh raises the total clean-fuel cost to 4.38 mills/kWh or 44.9¢/10^6 Btu coal or 57.7¢/10^6 Btu gas. Since the cost attributable to the coal itself in 20¢/10^6 Btu coal, that attri-

----

*This value is 4% less than that used in the UAC report because the latter does not account for the electrical generator efficiency (ca. 98%) or for the pumping load (2%)(Robson, 1970).

butable to gasification and cleaning is as shown in
Table 3.2. In comparing these costs with that of SO_2
removal from the stack gases of conventional steam
plants (Section 2.2), one must remember the thermodynamic cycle advantage obtainable by use of high-pressure
gas fuel in a combined gas turbine-steam turbine cycle
(Section 5.2).

Table 3-2. Gasification and Cleaning Costs for Lurgi-
Based System

Assumed Basis	¢ per 10^6 Btu in the		Mills per kWh electrical energy generated from gas at efficiency of	
	coal	gas	45%	38%
Sulfur credit taken	22.6	28.7	2.20	2.58
No sulfur credit taken	24.9	31.7	2.43	2.86

Analysis of the cost of shifting from use of a high-
pressure Lurgi producer to an atmospheric Wellman-Galusha
producer, using the same techniques of cost estimation,
indicates that atmospheric-pressure gasification adds
about 20% to the above costs.

Second-Generation Processes. Processes attainable in
ten years are referred to as second-generation and
assumed to be subject to a sulfur limitation in the producer gas of 90 g/10^6 Btu, equivalent to 200 ppm SO_2 in
the burned gases when 10%   excess air is used. The
most desirable second-generation processes employ direct gasification of pulverized raw coal with steam and
air in elevated-pressure, entrained-flow slagging gasifiers followed by gas purification, although some processes yielding valuable by-products would be attractive
if the by-products could be sold at current market
values. The dominant cost item in the first-generation
processes is the gasifier vessels. The conditions
specified above reduce the gasifier cost by yielding
higher reaction rates per unit volume of gasifier. Use
of oxygen instead of air would reduce equipment size,

but at a prohibitive cost.

As described earlier, elevated pressure in the gasifier results in lower gasification cost when the final gas is to be used in combined power cycles employing gas turbines. This is due partly to the higher specific reaction rates attained at elevated pressures and partly to the increase in gas volume accompanying gasification (volume of fuel gas is approximately twice the volume of air required for gasification). Elevated pressure would not necessarily be desirable in applications using low-Btu gas at atmospheric pressure, but it might be.

The rates of the gasification reactions increase strongly with temperature*; the required gasifier volume is consequently much smaller at higher temperatures above the ash fusion temperature (above about 2210 F). AS coal is heated to these high temperatures it tends to swell, soften, and sometimes agglomerate. This behavior, plus the slagging of the ash, makes fixed-bed operation extremely difficult, and favors entrained-flow gasifiers. Because of the small particle size, entrained-flow gasifiers are inherently cocurrent, and this minimizes tar formation.

A number of gasification schemes satisfy the above requirements. In each process air, steam and pulverized coal react in a cocurrent-flow gasifier, yielding mainly carbon monoxide and hydrogen. The chemistry of the different processes is essentially the same, but the gasifiers differ in construction detail and methods of solids handling. The gasifiers suggested are usually cylindrical, with reactants injected either axially

----

*The carbon-oxygen reaction in the combustion zone of the fixed-bed gasifier is a possible exception, but even there the primary combustion zone is very thin and does not control the volume required for fixed-bed gasification.

upward or downward, or tangentially, and with ash leaving as either solid or liquid slag. The coal is introduced into the pressurized system by lock hoppers, slurry
pumping, or a means not yet developed (e.g.,piston feeding, multiple lock hoppers with staged blowdown, etc.).

The cost attributable to the gasifier varies little
among the different proposed designs; the peripheral
equipment (heat exchanger, air compressors, scrubbers,
etc.) accounts for most of the cost. The gasification
reactions are quite rapid at temperatures between 2000 F
and 2500 F, and clean fuel gas containing no tars or un-
saturated hydrocarbons and negligible methane and higher hydrocarbons is produced. The gains from the faster
reaction rate at still higher temperatures do not compensate for the problems associated with operation at
such temperatures, and 2500 F is about the upper limit.

The higher temperature of cocurrent compared to
countercurrent solid and gas flow of course means a
higher ratio of sensible to chemical energy in the gas
leaving. About 20% of the energy of the hot fuel gas
from a cocurrent gasifier is sensible (versus less than
8% for conventional Lurgi). Recovery of most of this
heat in the fuel conversion process is necessary for
satisfactory process efficiency. Two possible alternatives for utilizing the sensible heat are (1) to generate high-pressure steam which can be expanded down to
gasification-process pressure in a noncondensing turbine
driving the air compressors and then be put back into
the process and (2) to preheat the gasifier reactants
(thus generating power only with burned gases). Without feed preheating, additional fuel must be oxidized
to provide the air for combustion. Consequently, the
required compressor power is increased and more fuel
energy appears as sensible heat in the product gases,
compounding the problem of heat recovery. The UAC
analysis (Robson et al., 1970), which considers these

factors as well as power-cycle efficiency, indicates
that the second alternative has lower overall costs.

Sulfur removal from the product gases can be achieved
by absorption in liquid scrubbing systems or on dry solids such as dolomite or iron oxide. Use of hot carbonate
scrubbing requires that the gases be cooled to 250 F,
while hot ferric oxide, for example, can be used on a
solid absorbent if the gases are cooled to below 1000 F.
Both options meet the second-generation sulfur emission
levels. Although the UAC report refers to commercial
use of hot ferric oxide, there may be significant problems associated with its use on a large scale and in
moving beds.

The Texaco Partial Oxidation Process, with a proposed
moving pebble-bed heat-recovery system and hot carbonate
scrubbing, illustrates the second-generation processes
(Figure 3-6). The gasifier has been operated pilot-
scale as a gas producer and commercially for making
synthesis gas (von Fredersdorff and Elliott, 1963).

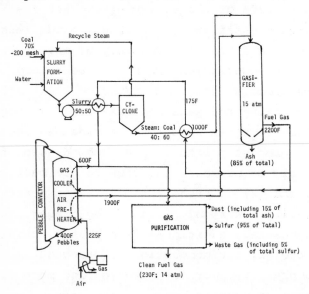

Figure 3-6. Clean Fuel Gas from Coal Using Texaco Partial
Oxidation Gasifier, Hot-Gas Heat Exchanger, and Gas Purification System (Robson, Giramonti, Lewis and Gruber, 1970)

Coal is crushed to 70% through 200 mesh, slurried with
an equal weight of water, and pumped to a suitable pressure for injection into the gasifier operating at 15 atm.
The slurry is vaporized by heat exchange with the hot
product gas and then adjusted to a 40:60 steam-coal
ratio in a cyclone; excess steam is recycled to the slurry tank. The steam-coal mixture is then preheated to
1000 F by a second heat exchange with the product gas.
(According to the Bureau of Mines the slurry feed system could have fouling problems (Holden et al., 1960).
Lock hoppers would be an alternative.)

Combustion air is compressed and preheated to 1900 F
in pebble-bed heat exchangers in which alumina pebbles
form a downward-moving bed. (Since the raw gas carries
about 15% of the ash, the operation of such heat exchangers at temperatures above the ash fusion temperature
is now questionable, but it may be possible in 10 years.)
The upper section of the beds is preheated to 2050 F by
a fuel gas entering at 2200 F, flowing countercurrently
to the pebbles and cooling to 600 F. Hot pebbles then
flow through a seal leg to the lower section of the exchanger, where they are cooled to 450 F and then recycled from the bottom of the bed to the top by a bucket elevator.

Preheated steam, coal and air flow into an 11.75-ft-
diameter refractory-lined gasifier 35.5 ft high. Approximately 95% of the coal is gasified during the 3-sec
residence time, and 5% is entrained with the slag and
lost. Fuel gas leaves the reactor at 2200 F, essentially in equilibrium with respect to the water-gas shift
reaction. It is estimated that about 85% of the coal
ash is trapped in the gasifier as slag which coats the
walls of the reactor and flows by gravity to a water-
filled pool at the reactor bottom. A finely divided
suspension of slag grains is removed. Hot fuel gas at
2200 F leaves the reactor about two-thirds of the way

down.   After preheating the reactants, the gas at 600 F
is available for dust and sulfur removal.   For a nominal
1000 MW plant, four trains of gasifier-scrubber-cleanup
equipment would be used.

The efficient use, <u>within</u> the gas-making system, of
the sensible heat of the gasifier gas gives this system
the high thermal efficiency, from coal to 230 F gas, of
87%.   Each pound of coal (MAF) reacts with 0.75 lb steam
and 2.6 lbs air, giving 73 SCF of fuel gas* containing
H_2S equivalent to 47 ppm SO_2 in flue gases.   These yields
are based on sulfur removal by hot carbonate scrubbers
followed by a Claus process.   While 95% sulfur removal
is indicated, 99% could be achieved at a higher cost.
The sulfur removal process also removes 75% of the CO_2.
Either elemental sulfur or sulfuric acid can be produced as a by-product, but the latter results in a 10%
higher effective fuel cost for the particular by-product credit rates assumed.

From the UAC cost estimates it can be inferred that
the cost of manufacturing gas with the above process,
exclusive of the coal cost itself, is 14.4¢/10^6 Btu if
a \$25/ton sulfur credit is taken, or 17.6¢ without sulfur
credit;** the associated plant capacity is 150 x 10^9
Btu/day (sensible plus chemical heat) of fuel at 206
psia (14 atm) and 230 F.   Based on coal at 30¢/10^6 Btu
and no sulfur credit, the gas would then cost 52.1¢/10^6
Btu, a figure not to be compared with conventional coal-
steam systems until allowance is made for the high cycle

----

*Higher and lower heating values are 179 and 162 Btu/SCF;
sensible heat above 60 F is 3.2 Btu/SCF.

**These figures were deduced from the following informa-
tion in the UAC report: cost attributable to gas plant
plus coal (after taking sulfur credit) for power plant
having 54.5% efficiency--2.35 mills/kWh; coal cost--
20¢/10^6 Btu; sulfur credit--0.2 mill/kWh.

efficiency possible with a gas turbine-steam turbine
system fed with pressurized fuel gas (Section 5.2.8) and
for the cost of SO_2 removal from stack gas in conventional steam-coal systems (Section 2.2).

Adaptation of Pipeline Gas Processes. Many of the processes described earlier for making pipeline-quality gas
from coal can be modified for low-Btu gas production by
substituting air for oxygen, operating at lower pressures, eliminating catalytic methanation, and introducing other changes in the peripheral equipment. Thus
Diehl and Glenn (1970) discuss a modified version of the
Bigas process and Schora (1971) describes a possible
design derived from the Hygas process. The intrinsic
features of the pipeline gas processes offer considerable potential for satisfying the requirements imposed
on modern gas producers.

3.2.3. Research Needs

In order to realize the full potential offered by low-Btu
gas as a fuel alternative, research and development are
needed to develop the necessary gasification equipment.
Research needs range from laboratory work, designed to
determine the basic gasification behavior of American
coals, to pilot-plant tests of different gasification
concepts, and finally to plant-scale demonstration of
equipment. Since the possible applications of producer
gas vary in character and scale, research is justified on
a range of process types, such as fixed-bed, fluidized-
bed and entrained flow. Fixed-bed processes are limited
to lump coal, and the permissible gas flow velocity (bed
area rating of 300 lbs coal/ft^2 hr at 20 atm) limits the
size of individual gasifiers. Large applications such as
power plants require costly multiple units. Furthermore,
as described earlier, fixed-bed processes produce tar and
have not been successful with caking coals at elevated
pressures. These difficulties suggest that, though research on other process types may be more fruitful in the
long run, present pilot-plant efforts to improve existing
equipment are warranted. The Bureau of Mines stirrer development for handling highly caking coals in Lurgi-type
producers deserves strong support, and should be expanded
to include solution of the dust-soot-tar problem. An examination of the Lurgi technique of handling tar from the
low-caking Continental coals shows that the tar is returned onto the coal feed. This probably increases the
chance of some methanation but it also increases the total tar handled. With highly caking American coals the
proper point of tar return to the gasifier deserves study.

There is only limited experience with fluidized-bed
gas producers, but such processes may well have future
importance. The Winkler process (Diehl and Glenn, 1970)
operating at atmospheric pressure has been employed in

large units using predominantly brown coal, but operation
at elevated pressure has been carried out only in pilot
plants with non-caking fuels. Fluidized-bed gasification
processes leave a high-carbon residue which requires separate utilization, thereby complicating the process economics; and ability to use caking coals at elevated pressure has not been established. Possible techniques for
overcoming the problems caused by caking coals are outlined in Section 3.1. There is also a technique of pro-
moting ash agglomeration without coal agglomeration which
appears to work in the Godel "Ignifluid" fluidized boiler
with inclined traveling grate (Squires, 1970). Such a
technique might also be effective for gasification. Development of these or other techniques would require
large-scale equipment because of the scale-up problem of
fluidized beds.

The high outlet gas temperature and associated high
reaction rate and minimized tar problems of entrained-
flow gasifiers using pulverized coal make entrained-flow
a promising process for low-Btu gas production. This
process offers considerable flexibility of size and operating pressure, and very large capacity units can be
built. Other advantages have already been described.
Since some of the pipeline-gas processes being considered
also employ entrained flow, it is clear that this is a
technique of wide application which deserves considerable
research effort.

There is some pilot-plant and commercial experience
with entrained-flow systems, and laboratory studies at
elevated pressures have been conducted in connection with
research on pipeline gas production at the USBM (Forney
et al., 1970) and at BCR (Glenn, 1970), both of which organizations are planning pilot-plant studies as noted in
Section 3.1 (though the Bureau of Mines proposal includes
the possibility of no longer employing an entrained flow

or cocurrent flow section). More research is needed at
both laboratory and pilot-plant scales, and some of the
areas to be studied are outlined below.

a.

Many of the research needs on gas producers are essentially the same as those outlined earlier for pipeline gas
production. Thus studies of gasification fundamentals,
physical behavior of coal and ash in gasifiers, and entrained flow are described in Section 3.1. In these areas,
research on making low-Btu gas should focus on the use
of air instead of oxygen and on lower pressures than
those of interest in making high-Btu gas.

b.

One of the most important research needs pertains to the
heat-exchange operation to be performed on the hot raw
gas prior to sulfur removal. Possible process arrangements include cooling the gas in a steam boiler, heat
exchange with the gasifier feed streams or with the
cleaned gas, or combinations of these. There is need for
high-temperature heat exchangers not requiring preliminary gas cleaning (mainly ash removal), or for techniques
not requiring preliminary gas cleaning at high temperatures. Operation at temperatures exceeding the ash fusion
temperature is desirable. Suggested heat exchanger
designs for such operations include moving pebble beds
(Robson et al., 1970) and liquid metal systems (Diehl and
Glenn, 1970), but both of these require further development. These and other techniques that appear feasible
should be studied to identify the best process arrangements for achieving heat recovery.

c.

Research should be conducted to develop techniques for
removing H_2S and COS from producer gas at the highest
possible temperature. Solid adsorbents must be resistant
to chemical and physical degradation. Calcined limestone
and dolomite offer considerable promise. The reaction of

these and other possibly suitable adsorbents with H_2S and
COS under the conditions encountered in gas production
should be studied, with emphasis on the kinetics of adsorption and desorption of the sulfur compounds and the
duration of adsorbent activity during cyclic operations.

3.3. Oil from Coal

3.3.1. Introduction

The slow disappearance of natural gas and oil from the
United States scene will be partly compensated by nuclear
power. The need for clean fossil fuel, however, will undoubtedly continue both for processing and for power production, though the latter can be expected to shift
slowly from present dominance by fossil fuel to ultimate
dominance by nuclear fuel. Failure to couple the expectation that nuclear fuel will in the long run supply our
power with the recognition that this cannot happen fast
enough to eliminate the need for clean fossil fuels now
is a real danger; inadequate emphasis on research to produce clean fossil fuels both for power and for processing
operations could be disastrous. Coal can produce synthetic pipeline gas (Section 3.1), cheap clean gas (Section
3.2), or oil (this Section). Production of coal liquids
can follow one of four general paths: (a) production of
a hydrogen-carbon monoxide mixture from coal, followed by
catalytic synthesis of hydrocarbons (variations of the
Fischer-Tropsch process); (b) solvent refining of coal
with minimum hydrogenation, yielding a very heavy ash-
and sulfur-free bottoms product; (c) pressure hydrogenation of coal from the approximate composition $CH_{0.75}$ to
nearer $CH_{1.25}$, using H_2 or H_2-CO mixtures (variations on
the Bergius or Pott-Broche processes; and (d) staged
pyrolysis to minimize the fractional residue. (Although
the last of these methods makes gas as well as liquids,
it will be retained in this section.)

3.3.2. Fischer-Tropsch Synthesis

The only process which produces liquid fuels from coal
on a commercial scale is the Sasol plant in the Republic
of South Africa. This plant uses the Fischer-Tropsch
process and is operated by the South African Coal, Oil
and Gas Corporation. The major products are gasoline,
diesel fuel, and waxes. The process makes products very
expensively by American standards, and is generally considered lacking in economic interest in the United States.
A technical description is included in the appendix to
this section.

3.3.3. Solvent-Refined Coal

If the only objective of coal treating is the production
of a clean fuel, then hydrogenation to produce solution
can be minimized and be followed by ash separation and
conversion of sulfur to removable form. Solvent refining
was initiated with the limited objective of producing a
low-cost antipollution alternative to residual oil and
natural gas for use under boilers. The original laboratory development of the chemical processes involved was
carried out by Spencer Chemical Corporation, and an economic evaluation of a full-scale plant has been prepared
for Pittsburgh and Midway Coal Mining Company by Stearns-
Roger Corporation (Kloepper et al., 1965; Pittsburgh and
Midway, 1970).

A process flow sheet is given in Figure 3-7a. Coal
from crushers (-1/8 in.) is slurried with anthracene-oil-
type solvent and 30 to 40 pounds of hydrogen per ton of
coal. The slurry is heated and passed to a high-pressure
flash vessel at a temperature such that the liquid is
filterable. The vapor stream from this stage is processed through a series of flash vessels at successively
lower pressure and temperature to separate various fractions for hydrogen recycling, phenol and cresylic acid
recovery, and acid gas removal.

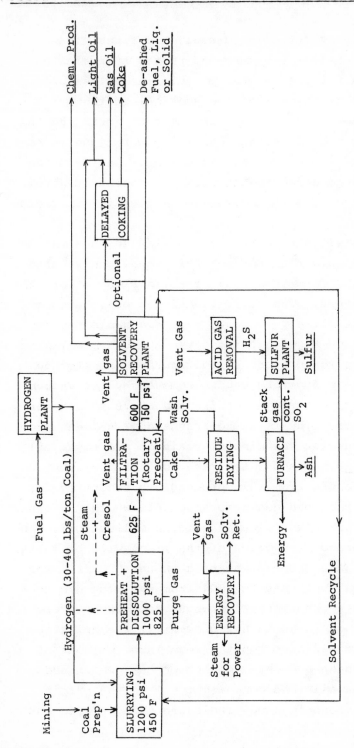

Figure 3-7a. Solvent Refining of Coal--Process for Making Ash-Free Sulfur-Free Fuel

The liquid portion of the dissolver effluent is
flashed to the filter pressure and passed to precoated
rotary filters for the removal of the mineral residue,
which includes nearly all of the ash, all of the pyritic
and half of the organic sulfur in the coal (bringing the
S below 1% for most American coals). The residue is sol-
vent-washed and stored for use as a fuel. Gas from the
filter is removed and combined with the condensate from
the vapor removed from the dissolver effluent, for treatment.

The liquid filtrate is heated and flashed in a vacuum
vessel. The liquid residue from this stage can be used
either in liquid form as a fuel or solidified to form the
final fuel product. The solidification process at commercial scale is likely to require considerable development, but Stearns-Roger has indicated the use of flaking
drums and silos for product solidification and storage.

The condensate from the vapors removed by the vacuum
flash stage passes through two fractionators to recover
various products. The first separates coal solvent from
the wash solvent for the mineral residues and light oil
products; the second separates wash solvent from the
light ends. Vapors from this process are recovered for
processing in the acid gas removal plant, while the final
liquid yields phenols, cresylic acid, and light oil. An
additional planned by-product of the plant will be sulfur,
coming from a Claus process running on the H_2S from the
acid gas removal unit and the SO_2 from the flue gas generated by burning the mineral residue from the filters.

A variation on the original process is the splitting
of the liquid de-ashed product stream into two, one of
which goes through a delayed coking process which adds
coke, coker gas oil, and more light oil to the original
products. An economic study indicates that delayed cok-

ing of up to 25% of the de-ashed stream improves the economic position of the process (Table 3-3). It is seen
that operation without coking permits sale of the de-
ashed, desulfurized fuel at 29.5¢/10^6 Btu (based on coal
from a plant-owned mine at \$4.15/ton), whereas operation
with 25% coking permits sale of the clean fuel at 19.5¢/
10^6. (These numbers assume close coupling of process
plant and power plant, and use of the fuel in liquid form.
If it is allowed to cool and solidify, about 2.3¢ additional cost is incurred). It must be pointed out, however, that this 10¢ saving by partial coking is the result of making the plant's chief income come from by-
product sales, which yield three and one-half times as
much income as the clean fuel product itself. If the by-
product credits are reduced 20% across the board, Table
3-3 indicates that the required selling price of the de-
ashed desulfurized product would be within one cent of
35¢/10^6 Btu, independent of what percent of the product
stream goes to delayed coking.

To assess the economic value of the clean fuel consider that, compared to the fuel feed cost of 16.2¢/10^6 Btu,
the clean-fuel product will cost an additional 3.3¢,
13.3¢, or 18.8¢, depending on whether the process operates with 25% coking, with 0% coking, or with 0% to 25%
coking but with by-product values reduced 20%. These incremental costs are to be compared to the cost of removing SO_2 from stack gases plus the cost of coping with the
ash problem in power plants. An average figure of 2
mills/kWh for SO_2 removal from stack gases may be used
for orientation (Section 2.2.5). This is equivalent to
22¢/10^6 Btu added fuel cost. In addition, Stearns-Roger
has estimated a saving of 0.36 mills/kWh consequent on
the lessened labor and power-plant capital cost of ash-
free operation. Tables 3-4 and 3-5 give their figures.
If the estimate of solvent-refining plant cost is realis-

Table 3-3. Economics of Making Solvent Refined Coal*(Pittsburgh and Midway, 1970)

	Without Coking	With 10% Coking	With 25% Coking
Investment Costs			
Mine	$23,555,000	$23,555,000	$23,555,000
Plant	71,381,000	79,632,000	86,648,000
Total	$94,936,000	$103,187,000	$110,203,000
Operating Costs			
Mine	$10,060,000	$10,060,000	$10,060,000
Plant	10,757,600	11,393,800	11,702,900
Total	$20,817,600	$21,453,800	$21,762,900
Cash Earnings Required for 10% DCF**Return	+14,400,000	+16,100,000	+17,500,000
Total	$35,217,600	$37,553,800	$39,262,900
By-product Credits	-16,930,000	-22,488,000	-30,861,000
Required Return from De-ashed Coal Product	$18,287,600	$15,065,800	$8,401,900
Total De-ashed Coal Product Available			
Tons/year	1,950,000	1,709,300	1,350,800
10^6 Btu/year	62,000,000	54,400,000	43,000,000
Required Selling Price of De-ashed Coal Product			
¢/10^6 Btu	29.5	27.7	19.5
[¢/10^6 Btu if By-Prod. C't in Error 20%]	35.0	36.0	33.8]

* Coal Consumption = 833,333 lbs/hr; De-ashed product in liquid state.

** Includes Depreciation.

Table 3-4. Economics of Conventional Power Station Fired with Solvent Refined Coal –
Capital Investment (800 MW) (Pittsburgh and Midway, 1970)

Item	Area Coal	De-Ashed Coal
Base Cost Installed Mechanical Equipment	$ 71,400,000	$ 71,400,000
Cost Deduction for:		
Fuel Handling Equipment	Base	$ 935,000
Ash Handling Equipment	Base	823,000
Soot Blowers, Piping and Controls	Base	460,000
Soot Blower Air Compressors	Base	520,000
Fly Ash Precipitator	Base	625,000
Smoke Stack and Breeching	Base	35,000
Steam Generator System	Base	1,935,000
Total Deductions		$ 5,333,000
Adjusted Cost, Installed Equipment	$ 71,400,000	$ 66,067,000
Process Piping, Fire Protection and Fencing	14,840,000	14,840,000
Electrical, Structural, and Remaining Plant	53,760,000	49,745,000
TOTAL CAPITAL INVESTMENT	$140,000,000	$130,652,000
Cost Per Rated kW at Generator Terminals	$175.00	$163.32

Table 3-5. Economics of Conventional Power Station Fired with Solvent Refined Coal –
Operating and Production Costs (Pittsburgh and Midway, 1970)

Item	Area Coal	De-Ashed Coal
Operating Labor (83/79 Men)	$ 705,500	$ 671,500
General Office (at 15 percent Area Coal Oper. Lab)	105,800	105,800
Payroll Overhead (at 25 percent of Total Payroll)	202,800	194,300
Contracted Maintenance (Labor and Material)	1,400,000	1,100,000
Water (7300 GPM at 9¢/10^3Gal.)	335,100	335,100
Treatment Chemical (at 6¢/10^3Gal.)	223,400	223,400
Miscellaneous Supplies (at 0.10 percent Capital Investment	140,000	131,000
Ash Disposal (at 60¢/ton)	80,500	N/A
SUB-TOTAL*	$ 3,193,100	$ 2,761,000

*Exclusive of land, site developments and fuel.

Table 3-5. Continued.

Item	Area Coal	De-Ashed Coal
Fixed Charges (percent Capital Investment)		
Interest on Investment	8.0	
Taxes (ad valorem, state, local)	1.5	
Insurance	1.0	
Depreciation	3.5	
Total Fixed Charges	14.0	
Annual Fixed Charges	$ 19,600,000	$ 18,291,300
TOTAL ANNUAL OPERATING COST	$ 22,793,000	$ 21,052,400
Annual Power Sales	5,355,000,000 kWh	5,397,245,000 kWh
Production Cost (exclusive of fuel)	$ 0.00426/kWh	$ 0.00390/kWh
Apparent Savings by Using De-Ashed Fuel		$ 0.00036/kWh

tic, the process is clearly economically interesting.
The small hydrogen consumption (30 to 40 pounds per ton
of coal fed), the absence of need for a catalyst, and the
simplicity that is associated with concentration on clean
fuel and minimized by-products together make this process
attractive and deserving of support through a pilot-plant
stage.

3.3.4. Consol Process

The Consolidation Coal Company has for many years been
developing processes for making synthetic liquid fuels
from coal, leading in 1963 to a contract with the Office
of Coal Research for design of a pilot plant. The plant,
dedicated in 1967, was intended to make liquids in the
gasoline range, and was christened Project Gasoline (Consolidation Coal Co., 1970). Later studies indicated the
advisability of changing the objective to the manufacture
of low-sulfur synthetic crude oil; and only the latter
process will be described here.

In the Consol process for synthetic crude production
Figure 3-7b), coal containing 2% moisture is crushed to
8-mesh or finer, slurried at 200 F with a process-derived
hydrogen-donor solvent in the ratio of 1 part coal to 2.5
parts of solvent, and heated to 700-800 F, where the coal
undergoes a complex dissolution and cracking process from
which gases and water are evolved. The remaining liquid
is a high-molecular-weight black oil having a melting
temperature around 400 F and consisting of solvent and
50-70% moisture-and-ash-free coal in solution, with un-
dissolved coal and ash in suspension. The black oil extract is cooled to 500-700 F and put through cyclone separators to form a relatively solids-free stream and a
solids concentrate. The latter is sent to a low-temperature carbonizer where it undergoes severe cracking to
produce char and synthetic distillates.

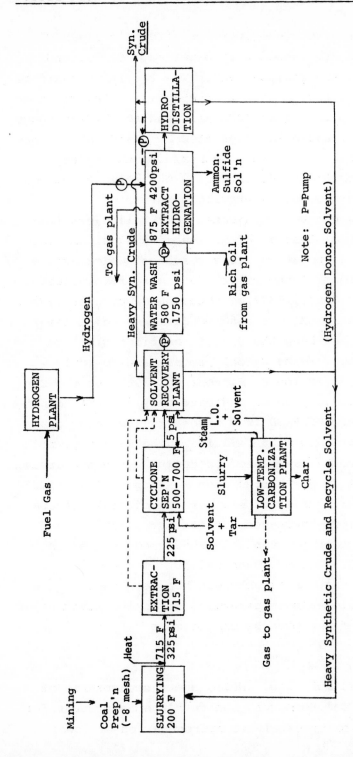

Figure 3-7b. Consol Synthetic Fuel Process--Synthetic Crude Production

The filtrate or cyclone overflow stream is flashed,
and the bottoms are washed with water at 1750 psi and
580 F to remove the residual ash. The resulting ash-free
extract is hydrogenated with zinc chloride catalyst in an
ebullated bed operating at 4200 psi, and the hydroproduct
is dropped in pressure and separated by distillation into
heavy recycle solvent bottoms (containing the hydrogen
donor) and synthetic crude product. The solvent bottoms
are returned to the coal-slurrying point.

This process has been subjected to detailed examination by a National Academy of Engineering Panel and the
Foster-Wheeler Corporation, as a result of which extensive modifications of the pilot plant were recommended
(Foster-Wheeler Corp., 1971). Funds have not been made
available to carry out the recommendations, and a later
recommendation has been the use of the pilot plant to
study a variation of the H-Coal process (see below).

Cost estimates (Ralph M. Parsons Co, 1969) on the Consol Synthetic Crude process indicate that a plant to produce 250,000 (46,000) B/SD from a Western (Eastern) coal
would cost about $633 (189) million. The larger plant
was estimated to be capable of producing synthetic crude
at a cost of $4.25 per barrel from Western coal at $1.25/
ton. Based on 3 barrels of oil per ton of coal, the barrel cost of oil is increased 33¢ for every dollar increase in coal cost per ton. In 1969 crude prices in
Wyoming were $2.85 to $3.25/barrel.

It is clear that until progress has been made on the
now-proposed pilot-plant operation, no real assessment of
the prospects of the process can be made.

3.3.5. H-Coal

Hydrocarbon Research, Inc. (HRI), under sponsorship of
the Office of Coal Research, has developed a process for
coal liquefaction by catalytic hydrogenation at lower

pressure than Project Gasoline. Crushed coal (Figure
3-7c) is mixed with recycle oil to form a slurry which is
pumped with hydrogen into a preheater operating at 2700
psi. The slurry plus preheated recycle gas from the main
reactor are pumped into the H-coal reactor, an ebullated-
catalyst column operating at 2700 psi and 850 F. The
catalyst, cobalt molybdate, settles below a point in the
bed at which liquid product is drawn off to a hot atmospheric flash drum. The product there separates into an
overhead stream, further separated by distillation, and a
bottoms stream, and the latter is split, part going to a
vacuum flash drum which separates it into vacuum overhead
product and bottoms slurry product, and part to a return
line to the slurrying operation. At the reactor the overhead vapors are partly condensed, and the uncondensed gas
(containing most of the fuel sulfur as H_2S) is sent to a
naphtha recovery operation, thence to acid gas removal,
and finally to the hydrogen plant with other fuel gas.
The flowsheet shows final products which must be subjected to further refinery operations. The char-oil product, containing unconverted solids, can be used as a fuel
or be subjected to carbonization to obtain more liquid
product.

The H-coal process has had bench-scale development in
a 3 tons/day process development unit. A proposal has
recently been made that a variation on the process, known
as the HRI fuel-oil process, be tested at pilot-plant
scale on the Cresap pilot plant of Consolidation Coal
Company,under contract of both companies to the Office of
Coal Research. The fuel-oil process will differ somewhat
from the one described above. A two-reactor two-stage
conversion system will be used, with the light and middle
distillate materials recycled with coal to yield the fuel-
oil product stream. Residual materials remaining unconverted would require separation and carbonization. It is

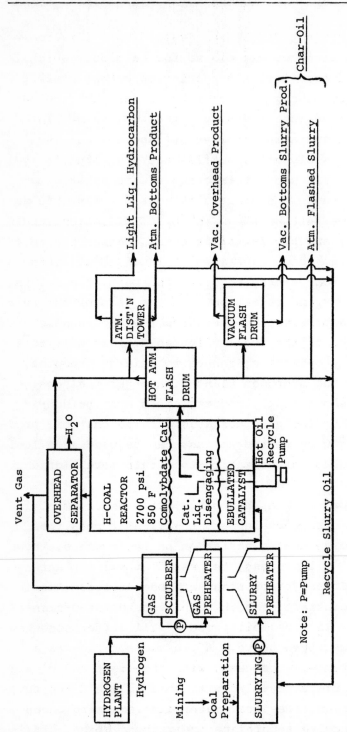

Figure 3-7c. H-Coal Process for Making Oil from Coal

expected (Office of Coal Research, 1971) that, after completion of negotiations, two years will be required to
prove out the H-coal fuel-oil process. A recent news
item (Chem. and Eng. News, 1971), unrelated to the above
plans, announces support of research on the H-coal process at a rate of $1 million per year for 1971 and 1972
by a group of five oil companies (Atlantic Richfield,
Continental, Humble, Gulf, Sun); plans for a prototype
plant are expected by mid-1972.

An economic projection for a 100,000 bbl/SD H-Coal-
based refinery has been made (Hydrocarbon Research, Inc.,
1968) in which the assumed yield of salable liquid products, based on data from the process development unit, is
3.5 to 4.0 bbl per ton of dry Illinois No. 6 coal or 2.8
to 3.2 bbl per ton of dry Wyoming sub-bituminous coal.
The principal products are gasoline and furnace oil (distillate fuel); several other products are also produced,
of which LPG and benzene are the most important. Costs
are based on 1967 values, including 12¢/gal for gasoline,
9¢/gal for furnace oil, and coal at $3.25/ton (Illinois)
and $1.20/ton (Wyoming). The projected return on the in-
vestment, which depends upon coal type, product spectrum,
and type of reformer used in hydrogen manufacture, is
given in Table 3-6 on the bases of book profit and total
cash flow. The return on investment as total cash flow
is 15.2% to 18.0% for Illinois coal and 14.3% to 16.4%
for Wyoming coal. These projections appear to be in reasonable agreement with another analysis of H-Coal (American Oil Co., 1967) which projects a 10% discounted-cash-
flow rate of return for conditions comparable to those
used above for Illinois coal.

3.3.6. Project Seacoke
The concept of fluidized-bed pyrolysis of coal with partial counterflow of gas and char to maximize the liquid

Table 3-6. Economic Summary for H-Coal [100,000 Bbl/SD Refinery (Hydrocarbon Research, Inc., 1968)]

Principal Products, 10^3 Bbl/SD				Type of Hydrogen* Reformer	Return on Investment	
Gasoline	Furnace Oil	LPG	Benzene		as Book Profit**	as Total Cash Flow+
Illinois No. 6 Bituminous Coal						
67	33	–	–	Nat. Gas	8.31	16.11
100	–	–	–	Nat. Gas	9.27	17.14
100	–	12	3	Nat. Gas	10.10	18.03
67	33	–	–	Coal & Char	7.49	15.23
100	–	–	–	Coal & Char	8.38	16.17
100	–	7.4	3	Coal & Char	9.06	16.91
Wyoming Subbituminous Coal						
67	33	–	–	Nat. Gas	6.88	14.57
100	–	–	–	Nat. Gas	7.87	15.63
100	–	12	3	Nat. Gas	8.55	16.36
67	33	–	–	Coal & Char	6.64	14.31
100	–	–	–	Coal & Char	7.53	15.27
100	–	12	3	Coal & Char	8.32	16.12

* Process makes enough gas for hydrogen manufacture except in third case from top. Reformer is fired with natural gas or coal and char.

** Book Profit = Profit – Profit – Taxes @ 48% of Profit, where Profit is Annual Revenue – Annual Expenses.

+ Total Cash Flow = Book Profit + Depreciation (15 year straight line) + Tax Credit (7% of Plant Investment Spread over Payout Period for Plant).

and gas yield from the VCM of the coal and produce gas,
liquid, and solid product streams, was developed by the
Atlantic Refining Company (now Atlantic Richfield Company)
in a process called Project Seacoke (ARCO, 1970). Another feature of the process was inclusion of petroleum
residium as part of the feed stock.

Figure 3-7d shows the process flowsheet. Crushed coal
containing approximately 2% free moisture is preheated to
approximately 310 F in the fluidized-bed preheater and
then pneumatically conveyed to the first fluidized-bed
reactor. Carbonization takes place in five reactors, so
arranged that the solids flow is in series, and with sol-

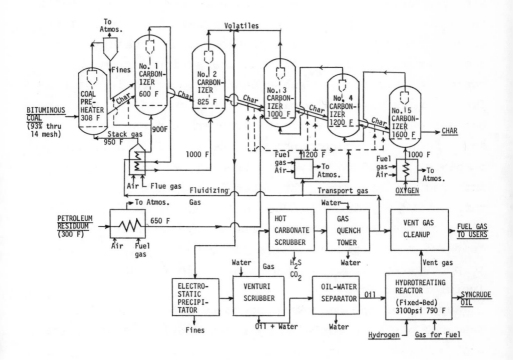

Figure 3-7d. Seacoke Process for Making Oil from Coal

ids temperature in each reactor controlled at a point
slightly below the initial plastic temperature. Backflow
of some of the char, 5 to 4, 4 to 3, and 3 to 2, is necessary to maintain a proper energy balance.

The gas streams flowing through carbonizers 1 and 2
and through 5, 4, and 3 join and pass through an electrostatic precipitator to remove entrained solids not removed in the cyclone system, thence to a Venturi scrubber. The gas is cooled to its saturation temperature of
approximately 185 F by evaporation of water, and any oil
mist is simultaneously removed. The mixture drains to
the oil-water settling basin. The gas stream at 185 F is
scrubbed with a carbonate solution to remove a portion of
the H_2S and CO_2, the former to eliminate the possibility
of an atmospheric pollution problem at the quench-water
cooling tower. The H_2S and CO_2 are stripped from the
carbonate solution and subsequently processed in the sulfur recovery unit. The oil-water mixture in the settling
basin separates into a light hydrocarbon phase, water,
and a heavy hydrocarbon phase. The two oil phases are
combined, heated to drive off free water, and pumped to
intermediate storage. The water is recycled to the Venturi scrubber. Crude product oil produced in the Seacoke
plant is reacted in the presence of hydrogen at elevated
temperature and pressure in a fixed-bed catalytic reactor
to produce Syncrude, a product suitable for use as a catalytic cracker feed.

Support of the Seacoke process by the Office of Coal
Research has been replaced by support of a rather similar
one, simpler in some respects--the COED process.

3.3.7. COED
This process--Char Oil Energy Development--developed by
F.M.C. Corporation is, like Seacoke, based on the multistage fluidized-bed pyrolysis of coal to produce oil, gas,
and char. Catalytic hydrotreating of the oil yields a

synthetic crude oil suitable as a petroleum refinery feedstock. The product gas can be re-formed to produce a
high Btu pipeline gas or hydrogen. The char product can
be used as a boiler fuel for power generation, or it can
be gasified to produce synthesis gas.

Figure 3-7e is an incomplete flowsheet. Pulverized
coal is fed through an air lock into two parallel trains
of equipment. Each includes a coal dryer, four fluidized
stages of pyrolysis, fluidized char cooler, and oil recovery and gas-recycle systems. The heat and gas required to dry the coal and to fluidize it in the first
stage are supplied by burning recycle gas from the oil
recovery system. Dried coal leaves the dryer at 375 F
and flows to the Stage 1 reactor. The bulk of the exit
gases from the dryer (N_2,CO_2,H_2O) is sent to or around
the first stage, and the remainder is vented. Exit gases

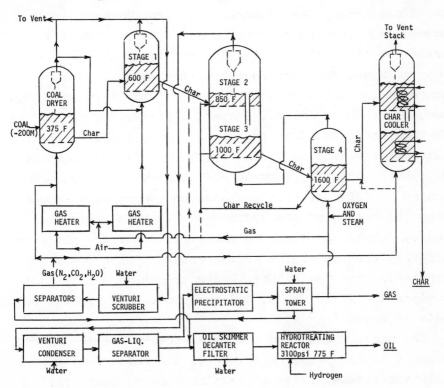

Figure 3-7e. COED Process for Making Oil from Coal

from Stage 1 (N_2, CO_2, H_2O) are Venturi-scrubbed and used
partly for fluidization in the char cooler and partly for
recycle to the dryer; and the oil and liquor from the
scrubber go to a skimmer-decanter system in the second-
stage recovery system.

Stages 2 and 3 are combined in one vessel. Product
gas and recycle char at 1600 F from Stage 4 supply the
heat required in the second-and third-stage reactors.
Product gases from Stage 2 flow to the oil recovery system. Product char from Stage 3 at 1000 F is heated to
1600 F in Stage 4 by combustion of a portion of the char
with oxygen. Product char from Stage 4 is cooled in a
fluidized-bed char cooler. The product gas from Stage 2
at 850 F passes through a Venturi condenser where it is
cooled to 170 F. Essentially all the oils are condensed
and removed in the gas-liquid separator. The effluent
gas flows through an electrostatic precipitator for fog
removal and then to a spray tower to remove the last
traces of oil. The gas leaving the tower at 100 F is
sent to a gas purification unit (not shown). The de-
canted oil, including that from the Stage 1 recovery system, flows to an oil dehydrator, a filter for removal of
char carried over from the second-stage reactor, and an
oil hydrotreating section. There the oil is pumped to
3100 psi, joined by recycle and make-up hydrogen, and
heated to 650 F by heat exchange on the product stream
from the bottom of the hydrotreater. This stream is
heated further to 775 F in a gas-fired furnace prior to
entering the top of the hydrotreater. Oil product is
separated from the lighter hydrocarbons in a series of
coolers and flash drums, and the product oil is pumped to
bulk storage.

The product gas from the oil-recovery section is compressed to 410 psia. H_2S and CO_2 are removed by a purification system, followed by a zinc oxide guard for removal of sulfur traces. The hydrocarbon gases are then

re-formed and shifted with steam at approximately 300
psia, and the CO_2 is removed. A methanation step then
follows for the final removal of CO. The product, available at 200 psig, is 95 volume-percent hydrogen, ca. 5%
CH_4. A portion of the COED product gas is used as process heat for the re-former section and for the other
areas where heat is required.

The COED process, like Seacoke, is intended to maximize the gas yield obtainable by coal pyrolysis alone,
with temperature staging to avoid agglomeration and countercurrent gas-char flow to minimize product decomposition. It succeeds in producing about the same char yield
as the standard ASTM proximate analysis for fixed carbon
plus ash. The process is stated (Jones, 1971) to have
produced, on a 30-day run on Colorado bituminous coal,
the following weight percent yields based on dry coal
feed (first column of numbers):

	Pilot Plant	Bench Scale (approx.)
Char	56.0%	60%
Oil	18.7%	20%
Gas	(9000 SCF/ton) 16.9%	(8000 SCF/ton) 15%
Gas Heating Value	-	535 Btu/SCF

The second column gives, for comparison, the results
of earlier bench-scale experiments. These and other (FMC
Corp., 1970) results, combined with product heating values, correspond to thermal efficiencies in the vicinity
of 100%. Such a high value is not realistic, and it is
not clear whether there were other thermal inputs; but
the data do support the reasonable conclusion that this
process operates at high thermal efficiency. The oil
yield of 18.7% corresponds to about 1.2 barrels of oil
per ton of coal.

Instead of use of the COED process to make the three
products listed above, the fuel gas (of about 500 Btu/
cu ft) can be used to make hydrogen at a claimed rate of
about 12,000 cu ft/ton coal for use in oil hydrogenation.
Similarly, the char can be used for synthesis-gas manufacture. With a scale of operation large enough to consider the gas streams from COED to be raw material for
pipeline quality gas, this process might be considered
for integration with the methanation operations of one of
the gas-making processes described in Section 3.1.

Economic studies (FMC Corp., 1970) of the process possibilities predate data acquisition on the pilot plant at
Princeton, New Jersey. Table 3-7 presents the results of
an economic analysis of the effect of shifting the gaseous product from fuel gas to hydrogen, and of using the
produced fuel gas versus char as the process fuel for the
operation. The table indicates that manufacture of hydrogen and oil, and use of 30% of the char produced to
supply needed process heat, results in the economically
most favorable operation. The assumption that coal and
oxygen costs are 25% higher reduces the percent return
(before taxes) by about 7 for all three combinations of
operation, but leaves the process still economically interesting. The analysis referred to will undoubtedly
soon be replaced by a better one based on pilot-plant
data.

3.3.8. Research and Development Needs--Oil from Coal

Valid recommendations of research expenditure on oil
from coal must rest on valid assessment of potential need
of oil from coal. Strong recommendations favoring federal support of pilot-plant research on pipeline-quality
gas from coal have been made here and by other groups as-
signed to study the problem. The supporting arguments
are these:

Table 3-7. Economics of COED Process (FMC Corp., 1970)

Case	I	II	III
Product	Hydrogen	Fuel Gas	Hydrogen
Process Fuel	Gas	Gas	Char
Fixed Capital, 10^6 $	36.17	32.70	44.65
Working Capital, 10^6 $-7%	2.53	2.29	3.13
Total Investment	38.70	34.99	47.78
Manufacturing Costs, 10^6 $/yr.			
Coal @ $3.00/ton (11.8¢/$10^6$ Btu)	10.50	10.50	10.50
Oxygen @ $6.00/ton	1.73	1.73	1.73
Utilities	1.70	1.39	2.36
Taxes, Insurance, Maint. & Depreciation	4.34	3.93	5.36
Labor, Supplies, PAC, & Overhead	1.20	1.20	1.44
Total Manufacturing Costs	19.47	18.75	21.39
Product Value, 10^6 $/yr.			
Char @ $2.75/ton	4.78	4.78	3.33
Oil @ $4.00/bbl.	18.80	18.80	18.80
Hydrogen @ 25¢/10^3 SCF	3.90	-	11.70
Fuel Gas @ 11.8¢/10^6 Btu	-	0.87	-
Total	27.48	24.45	33.83
ASR - 5% Sales	1.38	1.22	1.69
Net Sales Value	26.10	23.23	32.14
Gross Profit	6.63	4.48	10.75
Net Profit (50% Tax)	3.31	2.24	5.38
Percent Return (Before Taxes)	17.3	12.8	22.5
Discounted Cash Flow Rate of Return, %	13	10	16

a.

Energy requirements are rising rapidly, gas reserves are
dwindling, and the gas industry cannot look to gas manufacture from oil--a relatively easy and cheap process--
because the oil demands of a decade from now will be difficult to satisfy even without diversion of oil to gas.
b.

Federal control over gas rates through FPC gives the federal government a greater responsibility for the future
of gas than of oil, which is in the private domain (except for federal control through import restrictions);
and it is therefore particularly appropriate to sponsor
gas-from-coal research.

But if it turns out that marketing a clean Btu in oil
form from coal is markedly cheaper than marketing it in
gas form, the implications are far-reaching and of national significance. Federal support in the coal-to-oil
area becomes as important as in the coal-to-gas area;
there is no difference in the urgency of finding out
which fuel our industrial processing operations of the
future, and perhaps our space-heating operations as well,
should use in the interest of minimizing cost. On this
account a vigorous program of pilot-plant research on oil
from coal is recommended.

Of the four most promising processes described in this
section, two have pilot plants (Consol and COED), but one
of these is marking time without funds; one has the design for a pilot plant but no funds (Solvent Extraction)
and one is moving toward a pilot-plant design in 1972
(H-Coal). A National Academy of Engineering panel has
recommended action on the Consol process, Foster-Wheeler
has drawn up plans for modifications, and funds should be
made available to find whether the process is viable.
The Solvent Recovery process looks very attractive for
making ash- and S-free fuel for power plants; and support

of a project to build the pilot plant which has been designed is recommended.

When processes have reached the stage of those considered above, continued bench-scale research is warranted, but much of it is wasted without guidance from the
research output of the pilot plant. Bench-scale research
is needed on hydrocyclone design, filters, centrifuges,
furnace design, heat transfer in non-Newtonian liquids
containing suspended solids, physical ruggedness of catalysts, catalyst activity and revivification, solids separation, high-pressure vessel technology, and further
pressure reduction from the high pressures which characterized early Berginization.

3.3.9. Appendix (Fischer-Tropsch Process in Sasol Plant)

The Sasol plant gasifies coal at 400 psi in Lurgi reactors with pure oxygen and steam. Phenol and ammonium sulfate are produced as by-products of the gasification process. The raw gas from the Lurgi reactor is washed with
cold methanol to remove hydrogen sulfide, carbon dioxide,
and organic sulfur, then re-formed with high-purity oxygen and steam over a nickel catalyst to reduce the methane content. The gas from the re-forming section is preheated, mixed with recycle gas, and fed into the transfer
line, where it is mixed with powdered iron catalyst. The
catalyst-gas mixture flows through the transfer line into
the reactor, after which the catalyst is separated and
sent back. The reactor vapors are cooled to condense the
hydrocarbon products. Liquid hydrocarbons are treated to
remove oxygenated compounds, and distilled into gasoline
and fuel oil fractions.

Two types of reactors are presently being used, a Ger-
man-developed unit (Arge) which produces a mixture of hydrocarbons mostly in the diesel-fuel range, and a unit
designed by the Kellogg Company which operates at higher
temperatures and produces mainly gasoline. About 2000

tons/day MAF coal are gasified, and about 2400 tons/day
of coal are consumed by the power plant. The Sasol plant
produces 3660 barrels of 86-90 research octane number
gasoline and 370 barrels of diesel fuel per day. The
coal feed and products from the Arge and Kellogg units
are listed in Table 3-8.

3.4. Oil from Tar Sands and Oil Shale

The need for technological sophistication in recovering
oil from fossil deposits other than petroleum itself increases with the deposit type from tar sands to oil shale
to coal. The atomic hydrogen/carbon ratios of tar sands
and oil shale are also more favorable than that of coal
(ca. 1.5 for tar bitumen, 1.6 for shale oil, 0.72-0.92
for coal). These facts, plus the large magnitude of
known tar sand and oil shale deposits (Part I, this report), have stimulated much interest in those materials.

3.4.1. Tar Sands

Tar sands consist of agglomerates of sand particles each
in a thin envelope of water-suspended fines, with a high-
viscosity bituminous skin around the water envelope. The
bituman content of tar sands varies upward from 0 to
about 18%, with 14% considered rich. The total oil content of the world's tar sands is less than that of oil
shale, but it is impressively high. Spragins (1967) estimates 285×10^9 bbls to be recoverable from Alberta's
Athabasca deposit with current technology (for comparison,
oil-equivalent of U.S. total energy consumption in 1970 =
12×10^9 bbls). This deposit is currently being worked
by Great Canadian Oil Sands, Ltd. (GCOS) on a 45,000 bbl/
day schedule; and Syncrude Canada has Alberta government
approval to work an adjoining property which will be operable by 1976.

Recovery Methods. In-situ recovery of oil by underground
burning has appeared attractive because of its elimina-

Table 3-8. Sasol Coal Feed and Products (Gray, 1969)

Fuel Coal Analysis (Approximate)		Products (bbl/day) from	
		Arge Synthesis	Kellogg Synthesis
Moisture	6-8%	Gasoline 415	LPG 25
Ash	25-30	Kerosene 45	Gasoline (7 lb RVP) 3260
Volatile Matter	20-25	Diesel fuel 225	Diesel fuel 140
Fixed Carbon	40-45	Fuel oil 90	Waxy oil 45
Heating Value (LVH)	8000 Btu/lb	Wax 250	Methanol 15
			Ethanol 310
			Methyl ethyl ketone 22
			Acetone 16

tion of the enormous problem of handling overburden, sands and spent sands. Camp (1970) has reviewed the status of in-situ recovery. In the forward-combustion technique air flows from an injection well hole through heated spent sand to the combustion zone, the front of which moves away from the injection hole toward a production well hole along with combustion gases which sweep hydrocarbon vapors toward the production hole. In the reverse combustion technique air from the injection well flows through cold tar sands to the combustion front, which moves against the air into fresh sand; and the combustion products laden with hydrocarbons flow through heated sand to the production well hole. A modification of forward combustion, COFCAW (Combination of Forward Combustion and Water Flood), follows the forward combustion with air-plus-water injection to drive more bitumen to the recovery well. The vagaries of tar sand layer bounds, uncontrolled buoyancy forces and sand permeability and richness have so far combined to keep in-situ bitumen recovery from becoming commercial. Direct fluidized coking of tar sands has also apparently been abandoned (but see below, under "Research"). In contrast, the mining of tar sands followed by processing in a plant has generated much enthusiasm, one commercial plant, and a second one planned. The GCOS mining operation uses large bucket-wheel excavators, developed in German brown-coal strip-mining, to feed conveyors which carry the sand to the processing plant. One bucket wheel suffices for a 45,000 bbl/day operation, but two are used. Syncrude Canada, in contrast, proposes use of a large number of conventional scrapers to move the sands to the conveyor. Although a Stanford Research Institute study [(Roberts, 1970), see below] indicates that use of scrapers will yield about one percent higher rate of return on the total operation than use of bucket-wheel excavation, it is too early to make a fair comparison of the two mining methods.

Bitumen Recovery. The bitumen of tar sands is removed by
hot water treatment in a process of which the GCOS flowsheet (Figure 3-8) is typical. Hot water and sand, plus
some caustic, go to a conditioning drum, where sand lump
size is reduced by ablation, thence through a screen to
a separation chamber where three layers form. Sand is
withdrawn from the bottom and froth containing the bitumen from the top, and the middlings are partly recycled,
partly scavenged. The froth containing the bitumen
(about 88% of that in the feed) is ready for upgrading.
Upgrading. The bitumen is too viscous to pipe to a distant refinery. The GCOS upgrading process starts with
naphtha addition to the froth, followed by two-stage cen-
trifuging, coking and hydrodesulfurization to provide

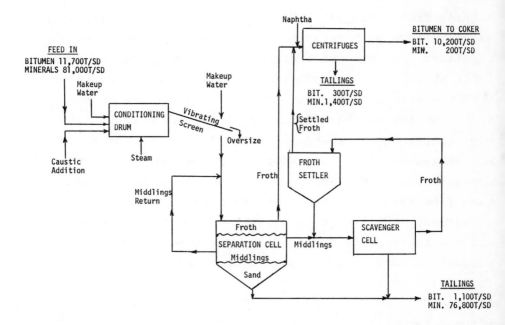

Figure 3-8. Great Canadian Oil Sands, Ltd. (GCOS) Process
for Bitumen Recovery from Tar Sands

both an upgraded product and the fuel needs of the plant.
The proposed Syncrude upgrading process is quite different and is based on H-Oil, a residuum hydrotreating process which starts with hydrovisbreaking of the bitumen.
The off-gas contains most of the sulfur as H_2S, and this
is recovered as sulfur. The liquid product is separated
into butanes, naphtha, light and heavy gas, oils, and
residue; and the middle three streams are separately hy-
drotreated catalytically.

The above descriptions focus on processes in or near
commercial operation. Roberts (1970) has made a detailed
economic analysis of the prospects for producing synthetic crude from tar sands, basing his estimate on a 50,000
B/SD (barrels per standard day; 350 days/yr) operation
using technology similar to that of the GCOS plant. The
figures from Table 3-9 include all on-site facilities but
exclude items such as access roads, townsite subsidy, and
pipelines. This consideration, plus GCOS operating experience which should enable the second builder to avoid
costly design changes, accounts for the difference between the sum of items 2 through 5 ($119.4 million) and
the reported GCOS expenditure of $270 million for a plant
of comparable size. The venture was evaluated, assuming
depreciation in fifteen years, by using the discounted
cash flow method to determine the rate of return on the
investment, which came to 5.8% when the value of crude
oil was taken as $2.90/bbl. The distribution of total
production costs according to processing areas is as
follows:

Mining		41%
Bitumen recovery		22
Upgrading	coke & S recovery 15	37
	hydrogen processing 22	
Total		100%

Table 3-9. Economics of 50,000 B/SD Tar Sands Plant (Roberts, 1970)

Required Investment (Millions)		Annual Cash Operating Costs and Income (Millions)	
1. Initial mine investment	$ 27.2	9. Mine operating cost	$10.9
2. Bitumen recovery	21.5	Plant Operating Cost:(10-15)	
3. Coker and sulfur recovery	28.8	10. Operating labor, supervision, overhead	3.2
4. Hydrogen & hydrotreating	30.4	11. Maintenance labor, overhead, and supplies	5.8
5. Offsite facilities	38.7	12. Catalyst and chemicals	2.6
Total initial depreciable investment	$146.6	13. Process royalty	0.5
6. Working capital	4.6	14. Plant overhead, taxes, insurance	3.6
7. Total mine replacement investment (15 yrs)	11.0	15. Alberta product royalty	5.1
8. Total start-tup expenses	15.8	Total operating costs	$31.7
		16. Syncrude income ($2.90/bbl)	50.8
		17. Sulfur income ($20/long ton)	2.3
		Total Annual Income	$53.1

The rate of return on the investment of course varies
with departure of plant capacity from 50,000 bbl/SD; with
departure of crude oil value from $2.90/bbl; with the
difference between onstream efficiency and 96 (corresponding to 350 days operation/yr); with difference between
percent bitumen in the sand feed and 12.3; with the difference between percent recovery of bitumen from the feed
and 88; and with difference between percent conversion of
recovered bitumen to synthetic crude oil and 75. The
formula below expresses the effect, on rate of return, of
small departures of the above variables from their standard values assumed in the main analysis:

$$
R = 5.8 + \left\{
\begin{array}{l}
2 \left(\dfrac{10^5}{50,000} - \dfrac{10^5}{\text{Cap'y, bbl/SD}} \right) \\[2ex]
5.7 \ (\text{Value, \$/bbl} - 2.90) \\[2ex]
0.23 \ (\% \text{ Onstream Eff.} - 96) \\
1.0 \ (\% \text{Bit. in sands} - 12.3) \\
.09 \ (\% \text{ Recovery} - 88) \\
.13 \ (\% \text{ Conversion} - 75)
\end{array}
\right\}
\qquad
\begin{array}{l}
\underline{\text{Range of Validity}} \\
25\text{-}150, 10^3 \text{bbl/day} \\[1ex]
2.80\text{-}3.30, \$/bbl \\[2ex]
70\text{-}100, \%\text{Eff.} \\
11.5\text{-}13, \ \%\text{Bit.} \\
75\text{-}95, \ \%\text{Rec.} \\
75\text{-}95, \%\text{Conv.}
\end{array}
$$

where R = rate of return on investment, %, by DCF method.
From the above relation it is clear that plant scale is
of decreasing importance above 50,000 to 100,000 barrels
capacity, that the rate of return will increase by 20% of
its value for a 20¢ per barrel increase in the price of
syncrude, or decrease by 20% for a drop in plant time on
stream from 96% to 91% or a drop in sand richness from
12.3 to 11.1; and that an increase in percent conversion
from 76 to 87 will add 1.5 to the 5.8% rate of return.
    Although the 5.8% rate of return is too low to be fi-
nancially attractive, the prospects of early increase of
that number are good, through oil price rise and through
improved mining and processing operations. It is fairly
clear that the period of 1930 to 1960, which saw a succes-

sion of unsuccessful small commercial ventures launched,
is at an end, and that world growth in energy consumption
has made the Athabasca sands a presently valuable energy
source. Factors unfavorable to tar sands, in any comparison of them with oil shale and coal as oil sources, include remoteness of Athabasca from markets (pipelining
oil 1500 miles to Chicago should add 45-70¢ to the barrel
cost), control of leases by the provincial government of
Alberta, U.S. oil import restrictions, and the severe
winter weather which could affect continuity of the enormous material-handling operation. Further comparison
with shale and coal will be made later.
Research. The existence of alternatives to the present
commercial process has already been mentioned (Camp,
1970). It is improbable that the next few commercial
processes will be near optimum because of the many points
in the process at which alternatives are available.
While decisions concerning these alternatives will be
made largely from pilot-scale experiments, opportunities
abound for bench-scale research by industrial, government,
and university laboratories. A few of the areas meriting
attention will be enumerated.
1.
In-situ thermal recovery. Two-dimensional modeling of
underground burning has received considerable attention,
but computer modeling has developed fast; and a reexamina-
tion of the comparison of physical modeling and mathematical modeling is in order. Mathematical modeling has its
shortcomings, but is much cheaper--and it is every day
coming closer to being able to match physical modeling.
Unanswered problems of economic minimum tar sand layer
thickness deserve reattack, using a good model.
2.
In-situ emulsion-steam drive. Shell Oil Co. (Doscher et
al., 1963), in a study of steam injection into tar sands,
concluded that injection of one-half ton of steam per

barrel of bitumen recovered would be required commercially. This corresponds to expenditure of one-sixth the
recovered bitumen as fuel to produce the steam. The process could have economic significance.

3.

Mining methods. Considering the enormity of the mining
operation associated with tar sand processing (110,000
tons sand and a comparable amount of overburden per day
for 50,000 bbl oil production), it is difficult to overemphasize the importance of mining machinery and mining
method development.

4.

Direct coking of tar sands. The Mines Branch of the Canadian Department of Mines and Technical Surveys has
studied direct fluidized-bed coking of tar sands. Two
fluidized beds are used, the first to accept the tar sand
and coke it, producing vapors which on partial condensation yield a synthetic crude oil stream and a gas stream
which is used to fluidize the coker and to carry hot
clean sand into it to supply the necessary heat. The
coked sand from the coker is carried by an air stream
into the second fluidized bed or burner bed, where flue
gas and 1400 F clean sand are produced. The ratio of
fresh tar sand to hot clean sand entering the coker is as
high as 5. As the process was developed, no practical
way was found for recovering the sensible heat in that
part of the hot clean sand stream which must be discarded,
amounting to about 10 percent of the energy in the bitumen entering the process. Direct fluidized coking is so
attractive a technique as to merit further research to
eliminate this defect. One possibility would be to use
the leaving clean sand stream to preheat the air for the
coker which could then be run rich to produce synthesis
gas which would be available to supply energy for other
parts of the upgrading process. Developments of this
type emphasize the need for background research on countercurrent heat transfer between sand and gas.

3.4.2. Shale Oil

Oil shale is a finely textured sedimentary rock containing the solid, largely insoluble organic material kerogen. High temperature decomposes the kerogen, yielding a raw oil suitable for use as a refinery feedstock. The principal deposits of oil shale are found in Colorado, Utah, and Wyoming. It has been estimated that these deposits contain two trillion barrels of oil, of which about 600 billion barrels are in rock assaying 25 gallons or more per ton (480 federal; 120 private).

There are two principal routes to recovering the oil contained in shale: conventional, above-ground retorting of mined shale, and in-situ retorting. Each approach involves several stages which have been developed to various degrees, as summarized in a Bureau of Mines chart (Figure 3-9).

Mining. Room-and-pillar mining for thinner, shallower seams has been demonstrated by the Bureau of Mines and TOSCO (The Oil Shale Corp., a consortium of oil companies). Open-pit mining for shallower seams is considered feasible but has not been demonstrated. It has also been suggested for the thick, deep seams for which no conventional mining technology has proved adequate, but to be economical such a mine would have to be enormous. "Cut and fill" mining is a second-generation approach to mining shale which the Bureau of Mines thinks might be suitable for the deep thick seams. Continuous mining machines remove the shale in layers, using spent shale as a floor on which to operate as higher levels of the shale are removed. This method offers the advantage of not leaving 50% of the resource in place as does room-and-pillar mining.

Retorting. The principal efforts at developing oil shale retorts have been made by the Bureau of Mines, Union Oil Co., and TOSCO.

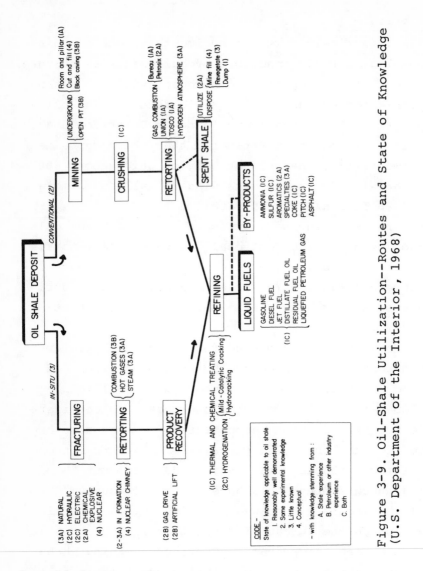

Figure 3-9. Oil-Shale Utilization--Routes and State of Knowledge
(U.S. Department of the Interior, 1968)

a.

In the BuMines Combustion Process crushed shale (1/4"-2")
is fed to the top of a cylindrical retort and falls successively through preheating, retorting, combustion, and
cooling zones to a grate and lock hopper. Recycle gas,
mostly inert but containing a small amount of hydrocarbon
vapor, flows up from the bottom. Air is introduced
through nozzles above the cooling zone, burning the residual carbon off the shale, producing a temperature of

1600-1800 F in the gas and on the lump surfaces, and de-
composing about one-quarter of the shale carbonate. In
the retorting zone where the kerogen decomposes the temperature is 800-900 F. Flow of the gases on upwards
through the fresh downcoming shale to the peripheral top
outlet ports cools the gas to 130 F and forms suspended
oil mist, which passes through centrifugal separators and
an electrostatic precipitator, yielding a crude oil of
20° API gravity and 80 F pour point and a gas which is
split into two streams, part being recycled to the retort and part going to the steam plant. The retorting
efficiency is 90% (the percent of that oil yield which
is obtainable in a standard Fisher Assay of the feed).
The process has been demonstrated on a 260 ton/day scale
by the Colorado School of Mines Research Institute.
b.
In the Union Oil process shale is charged into the lower
and smaller end of a vertical truncated cone and is
pushed upward against downflowing process gas. One retort in four uses air for partial combustion of the carbon residue in the shale (maximum temperature 2200 F);
and its product gas, after separation from the oil, is
burned to preheat the rich nitrogen-free recycle gas from
the other three retorts. The preheated recycle gas is
fed to the top of all four retorts, producing a maximum
temperature of 950 F in the three unfired ones. The process has the advantages over the BuMines process of not
dripping oil products back into hotter parts of the
charge for recracking, and of permitting an improved
thermal balance by proper portioning of the recycle gas.
The average retort efficiency (% Fisher Assay) is 91%.
The process was demonstrated at a rate of 1000 tons shale/
day in 1958.
c.
The TOSCO II retorting process uses hot ceramic balls to
heat the shale in a horizontal rotating kiln to 920 F

(Figure 3-10). The balls are separated from the spent shale, reheated in a furnace fired with product gas, and returned to the retort. Flue gas from the ball furnace is used to preheat the shale. The retort efficiency, based on standard Fisher Assay, is 105%. The process was demonstrated at a rate of 1000 tons/day in 1967.

Upgrading. The product of any of the retorting processes is too viscous for piping, and contains too much nitrogen

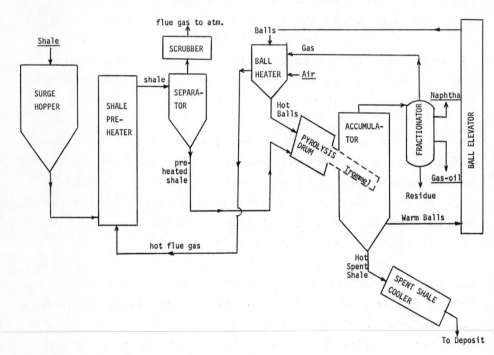

Figure 3-10. The Oil Shale Company (TOSCO) Process for Oil Shale Treatment

and sulfur to be used as a refinery feedstock. Figure
3-11, right half, shows the upgrading process, starting
with a 650 F flash into 40% overhead and 60% residuum
which is heated to 900 F and sent to delayed coke drums.
Gas oil and naphtha streams are separately hydrofined in
multiple-bed reactors consuming 2100 and 1500 scf, respectively, of 97% H_2 per barrel of charge. The final products, based on 50,000 bbl/SD of 43° API syncrude, are 820
T/SD of coke, 162 T/SD of ammonia, and 66 LT/SD of sulfur. The process consumes 92 x 10^6 cu ft H_2/SD, made by
re-forming natural gas with fuel from the process.

Process-Water Pollution. The water formed during the production of shale oil contains dissolved organic and inorganic materials which may or may not cause a major disposal problem. The amount of water formed and the amount
of contamination it carries, both of which depend upon the
type of shale and the retorting method, are of the order
20% to 40% of the oil produced (2% to 4% of the shale
processed) and 40 to 50 g/ℓ (Hubbard, 1971).* When water
is scarce, which is the case for much of the oil shale
land, the retort water would probably be reclaimed and
used, thus eliminating the disposal problem. Since some
operations will probably discharge water to the environment, and since water reclaimed for plant use will require
a certain amount of purification, the Bureau of Mines is
studying water treatment techniques, including combinations of (1) treatment with lime to remove carbonates,
most of the ammonia, and a portion of the organic materials, (2) adsorption on activated carbon to remove the
remaining organic material, and (3) ion exchange to re-

---

*Hubbard reports the following concentrations (gas-com-
bustion; in situ), in g/ℓ: Sodium (1.04; 3.10), Sulfate
(1.68; 4.45), Ammonium (8.91, 4.80), Carbonate (14.44,
19.22), Chloride (5.43; 13.41), pH (8.61; 8.69).

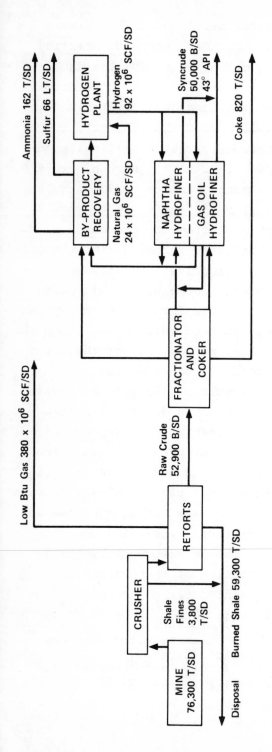

Figure 3-11. Syncrude from Oil Shale by the Gas Combustion Process (Murray, 1971)

move the balance of the cations and anions (Hubbard, 1971). This work, which is still in progress, is based on the use of materials that can be regenerated or discarded and on the recovery of by-products (e.g., ammonia and ammonia salts). Cost estimates of this treatment are not available.

Spent Shale Disposal and Runoff-Water Pollution. The spent shale from above-ground retorting, which amounts to 85% to 90% of the shale processed, is a source of pollution to surface streams and possibly to ground water. Water runoff from the waste piles contains the ions Na^+, Ca^{++}, Mg^{++}, $SO_4^=$, HCO_3^- leached from the spent shale. Analyses of runoff water from experimental shale residue piles subjected to simulated rainfall indicate contamination levels as high as 45 mg/ℓ at hydraulic equilibrium (when runoff rate equals rainfall rate)(Ward et al., 1971). A disposal technique suggested for eliminating the runoff problem, feasible only in some terrains, is belt transport to a narrow canyon provided with diversion ditches and culverts which prevent the spring runoff from entering the spent shale. Compacting, leveling, and planting when the final surface level is reached should prevent the development of a spreading eyesore. German brown-coal strip-mining practice deserves imitation. The difficulties associated with acceptable disposal in some terrains and the associated water runoff problem, however, constitute a major impetus to in-situ treatment of oil shale.

In-Situ Recovery. Recovery of oil by in-situ retorting of the shale involves fracturing the shale in order to create enough permeability to permit injection of retorting fluids, and recovery of shale oil through wells. The required fracturing is expected to create a large mass of broken shale ranging in size from dust to several feet and varying in assay value from nil to as much as 50 or

more gallons/ton. Heating is achieved by partial combustion of the fractured shale with injected air or a mixture of air and recirculated gas, or by some other technique.

Fracturing can be achieved with conventional techniques, of which hydraulic pressure, liquid chemical explosives, and high voltage electricity are being tested,
or by nuclear explosives. The latter technique, applicable primarily to relatively thick deposits under substantial overburden, is to be studied in Project Bronco.
The Department of the Interior estimates that a 200 kilo-
ton device exploded at a depth of 3000 ft would form a
420 ft diameter chimney of 7.3×10^6 tons of rubble containing 4.4×10^6 bbl oil of which 70% can be recovered.
Material outside the chimney is fractured to a smaller
extent, and oil recovery there requires higher retorting-
fluid pressures and achieves smaller recovery efficiency.
These estimations are quite uncertain since chimney size,
recovery efficiency, and required injection pressure and
flow rate of retorting fluid are not well known. Non-
nuclear techniques are expected to be applicable to both
thick, deep seams and to thinner deposits under shallow
overburden if the flow of retorting gases to the recovery
wells can be controlled. A more detailed description of
in-situ recovery is given by the Department of the Interior (1968).

In-situ processing has the advantages of (i) avoiding
the costs of mining and waste disposal, (ii) being adaptable to beds too thick, too badly faulted, or too deep to
be readily amenable to conventional mining methods, (iii)
eliminating or reducing needs for air-pollution controls
which would be required for above-ground processing, and
(iv) minimizing problems of possible contamination of
surface or subsurface waters by surface disposal of spent
shale wastes. Disadvantages are (a) in-situ processing

is not adaptable to concurrent recovery of associated
minerals, (b) fracturing methods are not yet developed,
(c) control of combustion may be very difficult, leading
to low recovery of shale oil, (d) the pressure at which
combustion air must be supplied is unknown, (e) the presence of faults may prevent combustion control.

Economics. Economic analyses have been made by the Bureau of Mines (Katell, 1967), by Hall and Yardumian (1968),
Lenhart (1969), and by Stanford Research Institute
(Murray, 1971). The last of these will be summarized
here (Table 3-10) because it is more recent and, in method, more nearly parallels that presented for tar sand
processing. Room-and-pillar mining and conveyor-belt
transportation of shale assaying 34 gal/ton were assumed,
under ideal conditions of shale access and disposal, with
no charge for shale other than that embodied in the initial mine investment. The assumed production rate was
50,000 bbl/SD of syncrude by the Gas Combustion process,
using twelve 36-foot diameter retorts (1970 technology).

The venture was evaluated, assuming 15-year depreciation and a depletion allowance of 15% of raw crude value
at the retort outlet, by using the discounted cash-flow
method to determine the rate of return on the investment,
which came to 9.9% for a syncrude value of $3.20/bbl.
Omission of coke, sulfur and ammonia credits drops this
to about 8.0%. Differences in the rate of return due to
use of the Union or TOSCO retorts instead of Gas Combustion were within the accuracy of the above estimate,
though there was some shift in the distribution of total
production costs for the TOSCO process--more efficient in
recovery, but retort more expensive. That distribution
according to processing area and retort type was as
follows:

	Gas Combustion	Union	TOSCO II
Mining	38%	37%	30%
Retorting	29	30	38
Upgrading	33	33	32

Table 3-10. Economics of 50,000 Bbl/SD Shale Oil Plant (Murray, 1971)

Required Investment (Millions)		Annual Cash Operating Costs and Income (Millions)	
1. Initial mine investment	$ 22.7	10. Mine operating cost	$13.05
2. Shale crushing and retorting	35.9	Plant Operating Costs:	
3. Coking and by-product recovery	14.4	11. Op. labor, and overhead	2.74
4. Hydrofining and hydrogen	27.5	12. Purch's'd mat'ls & utilities	7.74
5. Offsite and general facilities	21.8	13. Plant fixed costs	2.92
Total depreciable investment	$122.3	14. Process royalties	0.40
Noncapitalized Investment:		Total operating costs	$29.26
6. Land	$ 6.6	15. Syncrude income ($3.20/bbl)	55.50
7. Mine and plant working capital	10.8	16. Coke ($4.00/ton)	1.09
8. Total mine replacement		17. Sulfur ($22/long ton)	0.62
investment (15 years)	15.2	18. Ammonia ($25/ton)	1.36
9. Total start-up expenses	10.3	Total Annual Income	$58.57

The rate of return on the investment varies with departure of plant production rate from 50,000 bb/SD; with
departure of crude oil value from $3.20/bbl; with the
difference between onstream efficiency and 96 (350 days
operation/yr); with difference between shale assay and
34 gal/ton; with difference between shale disposal cost
and 5¢/ton. The formula below expresses the effect, on
rate of return, of small departures of the above operating variables from their standard value.

$$
\begin{array}{lll}
& & \text{Range of Validity} \\
\left[1.4 \left(\dfrac{10^5}{50,000} - \dfrac{10^5}{\text{cap'y, bbl/SD}} \right) \right. & & 30,000\text{-}100,000 \\
6.3 & (\text{Value, \$/bbl-3.20}) & 2.50\text{-}4.00 \\
R = 9.9 + \left\{ 0.2 \right. & (\text{Onstream Eff.-96}) & 85\text{-}100 \\
0.56 & (\text{Shale assay} - 34) & 23\text{-}35 \\
\left. 0.09 \right. & (5 - \text{Disposal Cost, ¢/ton}) & 0\text{-}30
\end{array}
$$

where R = rate of return on investment, %, by DCF method.
From the above it is clear that, as with the processing of
tar sands, plant scale is of decreasing importance above
50,000 to 100,000 barrels capacity. The rate of return
would drop from 9.9 to 8.0 if the cost of shale disposal
rose from the 5¢/ton assumed in this analysis to 26¢/ton.

The difference in syncrude value assumed between this
analysis and that on tar sand processing ($3.20 versus
$2.90) presumably reflects the differences in transporta-
tion cost from the Athabasca tar sand deposits and from
the Green River shale oil region to central U.S. markets.

The BuMines economics analysis (Katell, 1968) is somewhat more sanguine than the above SRI analysis, especially in allowing about 61¢, instead of 21¢, for by-products associated with 1 barrel of syncrude, in postula-
ting the feasibility of 45 ft retorts versus 36 ft, and

in estimating retort capacity on the basis of twice the
peak throughput per sq ft achieved in pilot-plant operation. The BuMines statistics support the statement that
above-ground retorting is cheaper than projected in-situ
operations, that open-pit mining will be cheaper than
room-and-pillar, and second-generation cut-and-fill still
cheaper.

Status of Development. The projected rate of return on an
oil shale processing plant in the Green River area using
1970 technology is greater than that on Athabasca tar
sands; but not as much progress has been made toward commercial operation. All the parts of the flowsheet have
been separately tested. Considered in reverse, they are:
a.
Upgrading. The separate items of this division are based
on well-known oil refinery technology. The optimum flowsheet has not been put together, but three different economic analyses support the claim that optimization in this
area will contribute in a relatively minor way to reduction in dollar cost per barrel.
b.
Retorting. The BuMines Gas Combustion retort has operated at 260 tons/day capacity, the Union Oil retort at
1000 tons/day and the TOSCO retort at 1000 tons/day. Extension to 36 foot diameter and an 8-fold increase in
throughput per retort, as projected in the SRI analysis
(Murray, 1971), appears to be achievable with little risk;
and the same analysis indicates the final total cost to
be rather sensitive to the choice among retorts.
c.
Mining. Despite much experience with mining of other
ores presenting mechanical problems similar to shale mining, experience with mining and acceptably disposing of
shale itself at a rate in excess of one ton per second is
missing. That is really the big hurdle. The many stud-

ies of large-scale mining methods, however, combined with
such experience as that of Union Oil and TOSCO at 1000
tons/day, point to the near certainty of being able to
keep the mining costs below 40% of the total.

The big question is mining and disposal, and there is
no substitute for full-scale experience. The venture, at
least the first one, may not be very profitable but it
can hardly be extremely costly. On the principle that
when there is need or impending need for very large-scale
production of an item (oil), the best method of making it
will not be found unless several competing methods are
vigorously pursued for enough years to allow time and opportunity for the flowsheet and management of each to become optimized-- on that principle the move toward early
commercial-scale production of oil from oil shale appears
warranted even if oil from coal is higher on the list.
Research Needs.   There follows a far-from-comprehensive
list:
1.
Mining. The chance of lowering the total cost of oil
from shale is so much more dependent on ore-handling methods than on any other step in the process that it is difficult to overemphasize the need for mining research.
Mining methods are well surveyed by Seegmiller and
Willson (1968). Research divides roughly into four categories: (a) Use of presently available mining equipment,
including studies of different ways of using a given kind
of equipment on a given geological formation, comparison
of use of different equipment on the same terrain, systems analyses to establish optimum location of refinery;
(b) invention of new equipment specific to different
types of oil shale deposits, including remote control underground mining machines; (c) consideration of novel
techniques, such as close-coupling an underground retort

with continuous mining machines (with oil-mist-laden gas
piped to a surface refinery); and (d) study of scale optimization of mining equipment.

2.

Oil Shale as a Mineral Source. The Department of the
Interior (1968) has pointed out the possibility both of
separating oil shale from other minerals lying next to
the oil shale deposit and of using spent shale as an
other-mineral source. The minerals of possible interest
are trona ($NaHCO_3 \cdot Na_2CO_3$), nahcolite ($NaHCO_3$), dawsonite
($NaAl(OH)_2CO_3$) halite ($NaCl$), shortite ($Na_2Ca(CO_3)_2$).
Trona is mined in the Green River Basin of Wyoming and
converted locally to soda ash. Studies of markets for
shale minerals, of their possible use as catalysts, and
of separation techniques are warranted.

3.

Heat Transfer. Heat transfer and pressure drop associated with gas flow through fixed beds of crushed shale
and through rotating inclined cylinders must be capable
of reasonably quantitative evaluation if oil shale retorting is to develop in the direction of increased thermal
efficiency through use of indirect heating of the retort
with gas heated in a separate chamber where only the carbon residue on the shale is burned. Such heat transfer
studies and flowsheet developments should be encouraged.

4.

Geologic Data. There is need for shale assays, bed inven-
tories, and geologic information basic to estimation of
mining costs, which together supply raw material for equi-
table federal-private land trades that consolidate hold-
ings and minimize lease troubles.

5.

In-Situ Recovery. This approach deserves exploratory
work, using mathematical and physical modeling or other
experimental techniques. Problems deserving attention

include three-dimensional propagation of burning and pyrolysis waves in porous solids, and pressure and flow rate
requirements for injected retorting fluids. Because suggested underground methods using nuclear explosives are
intrinsically quite inefficient, with the most sanguine
estimations of oil recovery inside and outside the chim-
neys being only modest and poor, respectively, development work in this area should be withheld until techniques which give promise of more encouraging results
have been projected.

3.5. Comparison of Proposed Fossil Fuel-to-Fuel Conversions

3.5.1. Introduction
Section 2 has presented the research and development status of the following conversion processes:

Pipeline-quality gas from coal
Cheap clean low-Btu gas from coal
Oil from coal, by{Solvent refining
                   Hydrotreating
                   Staged pyrolysis
Oil from tar sand
Oil from oil shale

It is appropriate now to ask the question, "How do these
compare?" But it is much more difficult to make the comparison than to develop reasons why a truly valid comparison is not presently feasible. The question will be
faced and comments and judgments will be made on the processes; but in the belief that the validity of the comments can better be assessed if the inherent difficulties
of comparison are first outlined, attention will be given
to the latter problem first.

There are at least seven considerations which make
intercomparison of processes difficult:
1.
The customer. For whom is the comparison made? Different interested parties will give different weight to the
different pros and cons of a process. For example, one
oil company may have a favorable position with respect to
rights to rich oil shale deposits with a thin overburden;
another may own tar pits which constitute a negligible
fraction of U.S. energy resources and not deserve national consideration but properly affect that company's views
on setting up a coal extraction plant. A utility company
and an oil company will tend to view the prospects of
clean ash-free heavy extract from coal differently. The
economic exploitation of the extraction process may involve, for power production at least, getting the power
company directly into a coal-processing plant operation;
for an oil company this may not be compatible with its
normal marketing practices. There are countless other
examples of company differences in resources and aims
which affect their relative weighting of two technologies
for making clean fuels.
2.
National energy policy. This has an enormous effect on
judgment as to when a process for fuel conversion will be
economic or as to which of two is better. If oil from
Athabasca tar sands is subject to import quota restrictions, the relative merits of Canadian tar sand extraction and Colorado oil shale retorting are affected. If
asphalt lakes in Venezuela are considered foreign oil
when the asphalt is hydrogenated before import, the dollar value of a large resource is subject to change by
congressional action. Government action with respect to
pollution control has a well-known profound effect on the
relative values of different fuels. If the sulfur re-

striction on power-plant stack gases is to change in 198x,
and x changes by federal action, the time at which a particular fuel-processing operation becomes economically
viable changes. Thermal water pollution restrictions
could make gas turbine-steam cycles more attractive and
thereby affect the economic position of clean low-Btu gas
from coal relative to another fuel which does not lend
itself so well to the power cycle postulated.
3.
Foreign energy policy and pricing of U.S. imports. The
record, during the past year, of oil import prices has
underlined the importance of this factor.
4.
Differences in status of development. Some of the processes described in earlier parts of this chapter have
not passed the bench-scale stage of development, a few
are based on pilot-plant data, several have acquired
pilot plants which have not yet yielded data. Comparative assessment under these circumstances is full of engineering guesses.
5.
Capacity to predict resources. We need to be reminded of
how dimly we see ahead, of how little allowance we made
for the effects of deep-well drilling, repressuring of
wells, offshore drilling, oil in Africa, gas in the North
Sea, Northern Slope oil--how little cognizance we had of
the impending effect of each of these, five years before
the event, on projected resources. We feel that at last
we have firm evidence that we know our gas and oil reserves rather well, and that a squeeze is almost upon us.
But we--or rather, an overlapping succession of us, each
with a span of ten to twenty years of responsible con-
trol--have felt that way several times before, with about
the same degree of confidence. Maybe we really do know
where we are now; but one wonders.

6.

Differences in procedures for cost estimation. Process
economic studies have been presented which fall into the
following rather widely different categories.
a. For pipeline gas-making operations, the rate of return is fixed at 7% and the analysis yields the price of
the product.
b. For power plants operating on clean specially-made
fuel, the analysis yields the cost of producing power,
using a power-plant-owned fuel-processing plant.
c. For production of clean fuel-for-power plus by-products, the analysis fixes the desired rate of return,
sells by-products at market values, and yields by difference the necessary cost of the clean fuel.
d. For oil from shale or tar sands or coal, the analysis
yields the rate of return on the process when all products are disposed of at market value.
These differences, combined with those due to differences
in methods used by different analysts for estimating capital costs of plants never built before, make comparisons
of processes difficult.
7.

Interchangeability of product fuels. It is almost axiomatic that natural gas is the easiest fuel to use. To the
combustion engineer it is almost axiomatic that where gas
is used some other fuel could be. In some cases the
equipment changes would be minor, in others major. For
small users of fuel or even for large users in which the
fuel cost is a minute contributor to total processing
cost, a plant may wisely choose to use its engineering
talent on problems critical to the viability of the process rather than direct them to the solution of fuel conversion problems, even if no equipment changes are necessary. The extent to which the natural-gas market is subject to shift of customers to other fuels, expressed as

a function of the relative Btu price of natural gas versus the other fuel, is simply not known. Under those
circumstances judgment on the importance of making gas
from coal at $1.00 per thousand cu ft versus synthetic
crude at $5 per barrel when the feedstock is 1000 miles
from the market is difficult. One could go as far as to
say that, on the above basis, the cost of one million Btu
of energy from these two fuels would be $1.00 and 86¢,
respectively; that the transportation costs would be
about 15¢ and 6¢, respectively; that the delivered energy
costs would therefore be $1.15 and 92¢, respectively, or
that oil would cost 80% as much per Btu as gas. But how
that would affect the market for the two is unknown.

With the limitations on validity of any comparisons
set out, the question arises as to what kinds of comparisons are appropriate. Clearly, the processes listed are
not all intercomparable. Comparison within three groups
appears logical: a consideration of clean power production; a comparison of oil from the three major nonpetroleum sources; a three-way comparison of pipeline-quality
gas from oil, oil from any source and, sometimes, low-
Btu gas locally made.

3.5.2. Clean Power from Coal

During the past year the shortcomings encountered with
sulfur dioxide cleanup of power-plant stack gases have
made processes for production of sulfur-free fuels or
processes for sulfur removal during combustion look especially attractive. The last of these methods has, un-
fortunately, received inadequate consideration in the
present study because of time and manpower limitations
(though see Section 2.2.4.). The remaining alternatives
are these:

Use of low-sulfur steam coals. This is only a short-time

solution because of limited supply; in addition, the premium on such coal is about $2/ton (ca. 0.8 mill/kWh).
Coal cleaning would be capable of lowering the sulfur
content to 1% in only about one quarter of the U.S. steam
coal (Section 2.2.3).

Use of low-sulfur Western coals. Large supplies of Western coals and lignites containing less than 1% sulfur are
available, but most power-plant furnaces have not been
designed for handling such fuels. More important, the
cost of transportation 1500 miles by unit train would be
about $9/ton (Section 2.1). Development of integral
trains would hopefully halve this number, but the cost
would still be high.

Use of low-sulfur oil. Only about 13% of U.S. central-
station power is based on oil. Most U.S. oil is high in
sulfur, the removal of which adds 50-80¢/bbl to the cost
(0.8-1.3 mills/kWh). The oil industry is investing heavily in sulfur removal plants, but the cost of removal from
residual oils is high. The U.S. oil consumption rate
presently exceeds the finding rate, and it is improbable
that fossil fuel's contribution to the anticipated doubling of power needs in ten years will come from sulfur-
free oil.

Reliance on present coal supplies, combined with stack-
gas cleanup. This has been found a more expensive operation than was claimed a few years ago, and stack-gas
treatment is expected to add $20 to $40/kW to power-plant
capital costs and 1 to 1.5 mills/kWh to operating expenses. The higher of the capital costs would add 0.8
mill/kWh to the power cost in a new installation, and
much more to that of a plant with less than half its useful life remaining.

Use of low-Btu clean gas from coal, in the gas-steam
cycle. In Section 5.2 the combination of a pressurized
gas producer with an advanced-cycle gas turbine followed
by a waste-heat boiler and steam turbine is shown to be
feasible now (German and Russian practice). A reasonably
realistic analysis of relative costs indicates that a
present-generation clean gas-steam plant (construction in
early 1970s) would produce power at only 1 mill more than
conventional use of coal, and that the former would require the addition of stack gas SO_2 treatment at about
2 mills additional power cost. The second-generation
gas-steam plant was estimated to produce power cleanly by
the 1980s for 1 mill less than a conventional coal plant
before SO_2 scrubbing is added.

Use of ash-free sulfur-free coal extract. It is pointed
out in Section 3.3.3 that if the objective in treating
coal is not to make simulated crude oil but to cause only
sufficient liquefaction to permit elimination of sulfur
and ash, the expensive hydrogen needed for liquefaction
is minimized. The product may be burned hot, in a power
plant close-coupled to the refinery plant, by atomization
in a manner completely analogous to smokeless burning of
tar and pitch, or it may be shipped cold in heavy-flake
form and burned as pulverized fuel. It is claimed that
the clean ash-free fuel can be sold for between 3.3 and
13.3¢/10^6 Btu above the fuel cost. Although this would
add 0.3 to 1.3 mills/kWh to power costs, the credit due
to absence of need for precipitators and ash-handling
amounts to 0.36 mill/kWh. Thus, even at the highest operating cost, labeled definitely off-optimum, the operation
is projected to cost less than scrubbing SO_2 from stack
gases.

The above six alternatives indicate clearly that two
concepts, making clean low-Btu gas from coal and making

an ash-free sulfur-free heavy hydrocarbon from fuel,
should be of great interest. If the cost estimates are
realistic--and they appear to be--completion of development of both processes should be supported. This calls,
in the first case, for federal funding of the coal extraction pilot plant which has been designed and, in the second, both for a more vigorous support of the work on coal
gasification with air and for the development of larger
advanced-cycle gas turbines.

3.5.3. Comparison of Coal, Tar Sands and Oil Shale as Oil Sources

All three of these sources of oil are so enormous that
they must certainly someday be tapped.* The timetable
for this, however, is very difficult to establish, especially because of changes in import quotas and in prices
established by foreign governments.

On a comparative basis tar sands appear not to have as
good a chance of early development as oil shale. Economic analyses of both processes have been made by the
same Institute (Murray, 1971; Roberts, 1970; and Section
3.4) on the same basis, and each process appears to have
been carried far enough in commercial or large pilot-
plant operation to make the analyses reasonably firm.
With synthetic crude priced at $3.60/barrel in Chicago,
the projected rates of return were 5.8% and 9.9% (discounted cash-flow method) for tar sands and oil shale,
respectively. In addition to this margin of shale over
sands, which probably exceeds the error in calculations
but which could disappear quickly as a result of a technological development, oil shale has the advantage that the
largest North American deposits are in the Green River

----
*Estimated Middle East oil reserves are about one-half
those of U.S. oil shale assaying in excess of 25 gal/ton
rock.

area (mostly Colorado) whereas the tar sands are mostly
in Alberta. It appears now that the processing of oil
shale in the United States is warranted earlier than extensive processing of tar sands, barring further significant relative technological change in the two processes
or the imposing of Western state restrictions on the mining of oil shale.

With respect to coal versus oil shale as sources of
synthetic crude oil the assessment is even more difficult.
Despite the considerable effort that has gone into oil-
from-coal research, processes for oil shale treatment are
simpler and more nearly ready for use. But the yield
from coal is so much greater (3 to 3.5 barrels/ton versus
0.8 barrel for relatively rich oil shale) and the disposal problem so much simpler that vigorous pursuance of
pilot-plant development followed by demonstration plant
operation is in the best national interest. The very fact
that the processing of coal to produce oil is considerably more complex than the production of oil from shale
is reason to expect a greater improvement, through research, in the efficiency and cost of the process.

The development of any of the previously discussed oil
recovery processes to the point of producing synthetic
crude at prices reasonably near present crude prices
could well turn out to be primarily an insurance policy,
but a very valuable one--insurance against excessive increase in the price of imported oil. On this ground
alone large federal support of fuel conversion research
is eminently warranted.

3.5.4. Gas from Coal versus Synthetic Oil from Coal
A technical affirmation popular today among engineers
knowledgeable in the energy field is this: We need to
learn how to make both high-Btu gas and oil from coal,
but gas from coal comes first because we are running out

of natural gas." Without denying the validity of the
conclusion, the next few paragraphs will examine briefly
the bases for it.

Data on annual oil and natural gas consumption, when
plotted on a logarithmic scale versus time, produce reasonably straight lines of markedly different slope. The
data on consumption and growth rate (the latter an average value, percent per year, from the identified year to
1970) are these:

Year	Oil Consumption, 10^{15} Btu/yr	% Growth per Year	Nat. gas Consumption, 10^{15} Btu/yr	% Growth per Year
1930	5.8	4.15	2.0	6.30
1950	13.6	3.95	6.0	6.87
1960	20.2	3.90	12.9	5.76
1970	29.6		22.6	

If these two growth rates stay constant at their 1960-
1970 average, the curves will cross at 55 x 10^{15} Btu/yr,
corresponding to 55 trillion cu ft gas/yr and 9.5 billion
barrels of oil/yr. No one is guilty of quite so naive a
prediction; the growth rates are generally assumed to
decrease. But conventional projections nonetheless tend
to be an extrapolation from the past into the future unmodified by considerations of cost and of interchangeability of energy types. A parallel projection of expected production of the natural products, gas and oil,
yields numbers which, subtracted from the projected total
U.S. requirements, give the synthetic gas and oil needs
of the country. These needs, arrived at substantially
without a consideration of anticipated costs of making
synthetic oil or gas, are sometimes presented as a basis
for decisions on needed research.

In the introduction to this subsection the question
was raised as to how energy consumers would react to the
equivalent of the hypothetical question, "Is oil at 80%
of the cost of pipeline gas a bargain?" Conversion of
hypothesis to fact would be very illuminating; it is not
quite possible today, but some preliminary thoughts can
be recorded. Estimated prices of pipeline-quality synthetic gas at the point of production (Section 3.1.5) appear below, with equivalent-Btu oil prices in parentheses:

Gas from Coal, 55¢ to 68¢/10^6 Btu ($3.19 to $3.95/bbl),
from coal at $4.60/ton.
Gas from Lignite, 47¢/10^6 Btu ($2.72/bbl), from lignite
at $2.25/ton.
But these figures were labeled optimistic, and the follow-
ing were offered as more probable:
Gas from Coal, 85¢ and $1 ($4.93 and $5.80/bbl)
Gas from Lignite, not given, but subject to similar escalation. These very iffy numbers may be compared with
those from Section 3.3 on oil from coal,
Consol, $3.25/bbl for S-free syncrude from Western coal
at $1.25/ton; $3.58/bbl for S-free syncrude from Western
coal at $2.25/ton
H-Coal, $3.78/bbl for S-free furnace oil from Illinois
No. 6 coal at $3.25/ton; DCF rate of return = 10%
COED, $4.00/bbl for S-free high-naphtha high-aromatic syn-
crude from coal at $3.00/ton; DCF rate of return = 13%,
10%, or 16%, depending on mode of operation;
and with those from Section 3.4 on S-free oil from sand
and shale,
Tar Sands, $3.60/bbl in Chicago; $2.90 in Alberta; DCF
rate of return = 5.8%
Oil Shale, $3.60/bbl in Chicago; $3.20 in Colorado; DCF
rate of return = 9.9%.

That the oil prices which were used in making these analyses are reasonably up to date is indicated by the fact
that July 1971 crude prices quoted in the Oil and Gas
Journal can be expressed by the relation
Price, $/bbl = ($3.25 to $3.48) + 0.02 (Gravity, °API-30)
The spread reflects primarily the effect of crude sour-
ness.

Comparison of the estimates of synthetic pipeline gas
costs with syncrude costs is especially difficult because
of the different accounting procedures used. FPC-regu-
lated gas prices are obtained by allowing 7% rate of return and finding the synthetic gas cost, which is of the
order of twice* the present cost of natural gas. Private-
industry syncrude prices are assumed the same as the market prices on crude oil, and the rate of return from the
operation is established. On the assumptions that sulfur-
free syncrude can be made by at least one process for
$3.60 with a 7% rate of return and that the best estimate
of synthetic pipeline gas price--in oil-equivalent--is
$5/bbl, one arrives at the conclusion that energy in syn-
thetic-oil form will cost 72% as much as in pipeline gas
form, before transportation. On the average, oil costs
about half as much to transport by pipeline as gas (Section 2.1); expressed as a price differential the figure
is 50¢/barrel per 1000 miles. (It is to be remembered
that all figures given are wholesale prices, and exclusive of distribution costs.)

It is clear that no claim can be made that the above
numerical comparison is known to be valid. A well-
grounded comparison will not be possible until pilot
plants for gas and for oil have led to demonstration
plants and the latter have been operated long enough to

---

*This is an oversimplification; natural gas is bought,
for shipment through pipelines, at 22¢ to 60¢/1000 cu ft.

have reached near-optimization.

Considering the billions of dollars per year hinging on the outcome of such a comparison, the need for large federal expenditures to develop clean synthetic fuels seems obvious.

It has not been the intention to imply in the above presentation that the cheaper fuel will win the race. Gas has advantages over oil which will offset a price difference (opposite in sign to the one existing today). As an aid to making more quantitative any comparison of gas with oil there is need for a comprehensive study of U.S. industry to determine the degree of interchangeability of fuel feasible at various cost differentials.

Related to the argument of the last several pages is the consideration of pipeline-quality gas versus low-Btu clean gas, the latter either made locally by the plant needing it or possibly considered for distribution from a central plant. According to Section 3.2, presently available Lurgi pressure gasifiers could make clean producer gas for 57.7¢/10^6 Btu. Since pipeline gas processes have not yet been developed, it is fairer to compare pipeline gas with what could be achieved in second-generation producers; the claimed cost would then be 52.1¢/10^6 Btu. This cost took into account the purchase of coal at 70% above the cost to a pipeline plant because of absence of mine ownership. It thus appears that clean producer gas might be able to supply industrial plants locally with gaseous fuel at a considerable saving over synthetic pipeline gas. This problem justifies a study similar to that on oil versus gas, to determine what markets would in the long run be better served by producer gas than by pipeline gas.

References

American Oil Co., 1967. "Evaluation of Project H-Coal," Research and Development Report No. 32, prepared for Office of Coal Research, Department of the Interior, Washington, D.C., December, 1967.

ARCO Chemical Company, 1970. "Project Seacoke," Research and Development Report No. 29, Final Report, Vols. I and II, prepared for Office of Coal Research, Department of the Interior, January, 1970.

Ballain, M. D. and Pulsifer, A. H., 1969. Paper presented at the 62nd Annual Meeting of A.I.Ch.E., Washington, D.C., November 16-20, 1969.

Benson, H. E., 1970. "Method and Apparatus for Producing Mixtures of Methane, Carbon Monoxide, and Hydrogen," U.S. Patent 3,503,724, to Consolidation Coal Co., March 31, 1970.

Camp, F. W., 1970. "The Tar Sands of Alberta, Canada," Cameron Engineers, Denver, Colorado.

Chemical and Engineering News, 1971. Vol. 49, No. 12, p. 59, March 22, 1971.

Consolidation Coal Company, 1970. "Summary Report on Project Gasoline," Research and Development Report No. 39, Vol. 1, prepared for Office of Coal Research, Department of the Interior, Washington, D.C., April, 1970.

Dent, F. J., Moignard, L. A., Eastwood, A. H., Blackburn, W. H., and Hebden, D., 1945. "An Investigation into the Catalytic Synthesis of Methane for Town Gas Manufacture," Gas Res. Board Comm., GRB 21/10, Great Britain.

Dent, F. J. and Hebden, D., 1949. "The Catalytic Synthesis of Methane as a Method of Enrichment in Town Gas Manufacture," Gas Res. Board Comm., GRB 51, Great Britain.

Department of the Interior, 1968. "Prospects for Oil Shale Development--Colorado, Utah and Wyoming," May, 1968.

Diehl, E. K. and Glenn, R. A., 1970. "Desulfurized Fuel from Coal by Implant Gasification," NAPCA Second International Conference on Fluidized Bed Combustion, Oxford, Ohio, October 4-7, 1970.

Dirksen, H. A. and Linden, H. R., 1960. Ind. Eng. Chem. 52, No. 7, p. 584.

Dirksen, H. A. and Linden, H. R., 1963. "Pipeline Gas
from Coal by Methanation of Synthesis Gas," Inst. Gas
Tech. Res. Bull. No. 31, July, 1963.

Doscher, T. M., Labelle, R. W., Swatsky, L. H., and
Zwicky, R. W., 1963. "Steam-Drive--A Process for In-Situ
Recovery of Oil from the Athabasca Oil Sands," in A Col-
lection of Papers Presented to K. A. Clark on the 75th
Anniversary of His Birthday, pp. 123-141, Research Coun-
cil of Alberta, Edmonton, Canada, October, 1963.

Field, J. H. and Forney, A. J., 1966. "High-Btu Gas via
Fluid-Bed Gasification of Caking Coal and Catalytic Meth-
anation," American Gas Association Synthetic Pipeline Gas
Symposium, Pittsburgh, November 15, 1966.

FMC Corporation, 1970. "Char Oil Energy Development,"
Research and Development Report No. 56, Interim Report
No. 1, prepared for Office of Coal Research, Department
of the Interior, Washington, D.C., May, 1970.

Forney, A. J., Denski, R. J., Bienstock, D., and Field,
J. H., 1965. "Recent Catalyst Developments in the Hot-
Gas Recycle Process," U.S. Bureau of Mines, Rept of Inv.
6609.

Forney, A. J., Gasior, S. J., Hayes, W. P., and Katell,
S., 1970. "A Process to Make High-Btu Gas from Coal,"
Bureau of Mines Technical Progress Report 24, April, 1970.

Foster Wheeler Corporation, 1971. "Engineering Evaluation of Project Gasoline Synthetic Fuel Process," Research and Development Report No. 59, prepared for Office
of Coal Research, Department of the Interior, Washington,
D.C., January, 1971.

Gary, J. H., 1969. "Liquid Fuels and Chemicals from
Coal," Mineral Industries Bulletin, Colorado School of
Mines, Vol. 12, No. 5, September, 1969.

Gilkeson, M. M., White, R. R., and Sliepcevich, C. M.,
1953. Ind. Eng. Chem. 45, No. 2, p. 460.

Glenn, R. A., 1970. "Status of the BCR Two-Stage Super-
Pressure Process," Third American Gas Association Synthetic Pipeline Gas Symposium, Chicago, November 17-18.

Hall, R. N. and Yardumian, L. H., 1968. "The Economics
of Commercial Shale Oil Production by the TOSCO II Process," 61st Annual A.E.Ch.E. Meeting, Los Angeles,
December, 1968.

Haynes, W. P., Elliot, J. J., Youngblood, A. J., and
Forney, A. J., 1970. "Operation of a Sprayed Raney
Nickel Tube Wall Reactor for Production of a High-Btu
Gas," Joint Meeting of the Division of Fuel Chemistry and
Division of Petroleum Chemistry of the American Chemical
Society, Chicago, September 9, 1970.

Henry, J. P., Jr., and Louks, B. M., 1970. "An Economic
Comparison of Processes for Producing Pipeline Gas
(Methane) from Coal," ACS Meeting, Chicago, September, 1970.

Holden, J. H., and associates, 1960. "Operation of Pressurized Gasification Pilot Plant Using Pulverized Coal
and Oxygen," Bureau of Mines Report 5513.

Hubbard, A. B., 1971. "Method for Reclaiming Water from
Oil-Shale Processing," American Chemical Society, Division
of Fuel Chemistry Preprints 15, No. 1, 21-25, March-
April, 1971.

Hydrocarbon Research, Inc., 1968. "Project H-Coal Report
on Process Development," Research and Development Report
No. 26, prepared for Office of Coal Research, Department
of the Interior, Washington, D.C., November, 1968.

JANAF Thermochemical Tables. Report PB-168-370 and Ad-
denda PB-168-370-1 and PB-168-370-2, the Dow Chemical
Company, Midland, Michigan.

Jones, J. F., 1971. Discussion of Project COED at Office
of Coal Research Contractors Meeting, Illinois Institute
of Technology, Chicago, May, 1971.

Katell, S., 1967. "A Cost Analysis of an Oil Shale In-
stallation in Colorado," circa 1966, Government exhibit
G-667 presented in Bureau of Land Management Colorado
Contests 359-360, U.S. Department of the Interior.

Kavlick, V. J., Lee, B. S., and Schora, F. C., 1970.
"Electrothermal Coal Char Gasification," Third Joint
Meeting of the Istituto de Ingenieros Quimicous de Puerto
Rico and the A.I.Ch.E., San Juan, Puerto Rico, May 17-20.

Kloepper, D. L., Rogers, T. F., Wright, C. H., and Bull,
W. C., 1965. "Solvent Processing of Coal to Produce a
De-Ashed Product," Research and Development Report No. 9,
prepared for Office of Coal Research, Department of the
Interior, Washington, D.C., by Spencer Chemical Division,
Gulf Oil Corporation, December, 1965.

Lee, A. L. and Feldkirchner, H. L., 1970. "Methanation for Coal Hydrogasification," Joint Meeting of Division of Fuel Chemistry and Division of Petroleum Chemistry of the American Chemical Society, Chicago, September 9, 1970.

Lee, B. S., 1970. "The Status of the Hygas Program," Third American Gas Association Synthetic Pipeline Gas Symposium, Chicago, November 17-18, 1970.

Lenhart, A. F., 1969. "The TOSCO Process-Economic Sensitivity to the Variables of Production," American Petroleum Institute Proceedings--Refining Division, pp. 907-924.

Mills, G. A., 1970. "Progress in Gasification--U.S. Bureau of Mines, Third American Gas Association Synthetic Pipeline Gas Symposium, November 17-18, Chicago.

Murray, R. G., 1971. "Economic Factors in the Production of Shale Oil," 74th National Western Mining Conference, Denver, February, 1971.

Nelson, H. W., Layne, H. N., and Hein, G. M., 1966. "Final Report on Study of Costs of Production and Potentia. Future Markets for (Phase I) Low-Btu Industrial Fuel Gas (Producer Gas) and (Phase II) Industrial Hydrogen," No. PB-174-835, to Office of Coal Research, U.S. Dept. of the Interior, from Battelle Memorial Institute, Columbus, Ohio.

Office of Coal Research, 1971. Annual Report, Office of Coal Research, Department of the Interior, Washington, D.C.

Pittsburgh and Midway Coal Mining Company, 1970. "Economics of a Process to Produce Ashless, Low-Sulfur Fuel from Coal," Research and Development Report No. 53, Interim Report No. 1, prepared for Office of Coal Research, Department of the Interior, Washington, D.C., June, 1970.

Ralph M. Parsons Company, 1969. "Feasibility Report Consol Synthetic Fuel Process Synthetic Crude Production," Research and Development Report No. 45, Interim Report No. 2, prepared for Office of Coal Research, Department of the Interior, Washington, D.C., July, 1969.

Roberts, P. V., 1970. "Comparative Economics of Tar Sands Conversion Processes," American Chemical Society Meeting, Chicago, September, 1970.

Robson, F. L., Giramonti, A. J., Lewis, G. P., and Gruber, G., 1970. "Technological and Economic Feasibility of Advanced Power Cycles and Methods of Producing Nonpolluting Fuels for Utility Power Systems," United Aircraft Research Laboratories, Report prepared for National Air Pollution Control Administration, U.S. Dept. of Health, Education, and Welfare, December, 1970.

Robson, F. L., 1971. United Aircraft Research Laboratories, East Hartford, Conn., personal communication, May.

Rudolph, P. F. H., 1970a. "New Fossil-Fueled Power Plant Process Based on Lurgi Pressure Gasification of Coal," American Chemical Society Division of Fuel Chemistry Preprints 14, No. 2, pp. 13-38, May, 1970.

Rudolph, P. F. H., 1970b. Personal communication with Robson, Giramonti, Lewis, and Gruber (1970), June, 1970.

Schlesinger, M. D., Demester, J. J., and Greyson, M., 1956. Ind. Eng. Chem. 48, No. 1, p. 68.

Schora, F. C., 1971. Institute of Gas Technology, Chicago, personal communication, May, 1971.

Seegmiller, B. L. and Willson, J. E., 1968. "Mining of Fossil Hydrocarbons," in Fossil Hydrocarbon and Mineral Processing, Chemical Engineering Progress Symposium Series 64, No. 85, A.I.Ch.E., pp. 51-56.

Spragins, F. K., 1967. "Mining at Athabasca--A New Approach to Oil Production," J. Petroleum Technology, pp. 1337-1343, October, 1967.

Squires, A. M., 1970. "Clean Power from Coal," Science 169, No. 3948, pp. 821-828.

Tsaros, C. L. and Joyce, T. J., 1968. "Comparative Economics of Pipeline Gas from Coal Processes," Second American Gas Association Synthetic Pipeline Gas Symposium, Pittsburgh, November 22, 1968.

Tsaros, C.L., 1971. Personal communication.

von Fredersdorff, C. G. and Elliott, M. A., 1963. "Coal Gasification," Chapter 20 in Chemistry of Coal Utilization, Supplementary Volume, (H. H. Lowry, Ed.), John Wiley and Sons, p. 982.

Wainwright, H. W., Egleson, G. C., and Brock, C. M., 1954. "Laboratory-Scale Investigation of Catalytic Conversion of Synthesis Gas to Methane," U.S. Bureau of Mines Rept. of Inv. 5046.

Ward, J. C., Margheim, G. A., and Löf, G.O.G., 1971. "Water Pollution Potential of Spent Oil Shale Residues from Above-Ground Retorting," American Chemical Society Division of Fuel Chemistry Preprints 15, No. 1, pp. 13-20, March-April, 1971.

Wellman, P., 1971. U.S. Bureau of Mines, Energy Research Center, Process Evaluation Group, Morgantown, W. Va., personal communication, February, 1971.

Chapter 4

NUCLEAR POWER

4.1. Introduction

Nuclear fission is an alternative to fossil fuel combustion as an energy source for central-station power generation. Since about one quarter (24.66% in 1970) of the U. S. energy goes into the production of electrical power by utilities, the potential importance of nuclear power is high. Nuclear reactors in 1970 processed 0.3% of the total energy consumed in the U. S. and supplied 1.2% of the utility power. At the end of 1970 a total of 20 operable nuclear plants constituted 2.2% of the electric utility capacity, and 89 additional plants were either being built or planned.

Light-water-moderated reactors--the kind used predominantly in the U.S. and in most other countries-- produce most of their power by slow-neutron fission of the scarce isotope of uranium, U^{235}, which is only 1/140 of natural uranium. The rest is U^{238}, which is capable of conversion by neutron absorption to the fissionable plutonium isotope Pu^{239}. Although some conversion of U^{238} to Pu^{239} occurs in light-water-moderated reactors, in a fast-neutron reactor it is possible to produce an amount of Pu^{239} from U^{238} that exceeds the amount of Pu^{239} used in fission. Thus the so-called fast breeder reactors would permit a larger use, and therefore an effective expansion, of uranium reserves. This possibility has stimulated great interest in the eventual replacement of light-water reactors with breeders. Since light-water reactors are now established on the energy scene and a large research and development effort would be required to bring breeder reactors to a position of commercial importance, the proper time scale for breeder development depends upon the cost-supply relationship of uranium reserves. A limited assessment of uranium reserves and of the status and potential of light-water and breeder reactors is given below.

4.2. Nuclear-Fuel Reserves

The Atomic Energy Commission (1970) gives the following
estimates of the U.S. uranium reserves in deposits known
or expected to be found, as a function of uranium price:

Price of Uranium Concentrates, $/lb U_3O_8	Tons of Uranium Resources at this or lower price
8	594,000
10	940,000
15	1,450,000
30	2,240,000
50	10,000,000
100	25,000,000

These reserves may be compared with the projected requirements for power generation which, based on use of light-
water reactors, are given in Table 4-1.

Table 4-1. Uranium Requirements for Supplying Projected
Nuclear Power Needs with Light-Water Reactors (Benedict,
1971)

Year	Electric Generating Capacity, 10^3 MW		Cumulative Consumption of Uranium Ore Concentrates (U_3O_8), tons*
	Total	Nuclear	
1970 (Actual)	300	6	---
1980	523	145	200,000
2000	1550	735	1,600,000

*171 tons of U_3O_8 generates 1000 MW-years of electrical
energy in light-water reactors.

According to the reserve estimates, the 1,600,000 tons
of uranium concentrates which would be consumed in light-
water reactors by the year 2000 would raise the price of
uranium from the present value of $8/lb to about $18/lb.

Since the cost of electricity from light-water reactors
increases by about 0.06 mill/kWh per \$1/lb increase in
the price of U_3O_8 (Benedict, 1971), the above increase
would add about 0.6 mill/kWh to the cost of electricity--
a 7% increase in busbar power cost. A much lower percentage would apply to both industrial and residential
customer prices owing to fixed transmission and distribution costs.

The use of fast-breeder reactors would increase
uranium reserves about 130-fold and make the cost of
electricity almost independent of uranium costs.

The above information leaves little doubt that uranium reserves are adequate to supply the projected demand
for nuclear power, and that increased fuel costs will
eventually occur if breeders are not developed. With
respect to the urgency for their development, however,
there is room for much disagreement. The history of
fuel-reserve projections is rich with underestimations,
and uranium projections may not be exceptional. Uranium
prospecting was vigorous a decade ago, but light-water
reactors have not become available as fast as was anticipated and the pressure to find new ore is reduced.
The state and completeness of our prospecting is perhaps
illustrated by the fact that a uranium discovery in
Australia last year may be larger than the known U. S.
supply. Such events underline the "iffy" character of
the AEC projection of price vs tonnage. Prospecting for
certain types of uranium deposits such as pitchblende in
igneous rock faults has not received the attention it
deserves. The mining industry would be enormously stimulated to search-and-find if the price of crude U_3O_8

were, for example, increased fourfold* in association
with extended use of light water reactors. Until there
is a clear cost advantage of breeders over light-water
reactors, independent of projected increases in uranium
cost, the extent of the need for hurried breeder development is no better established than is the knowledge
of uranium reserves.

4.3. Present Technology

4.3.1. Introduction

The two types of nuclear reactors now commercially
available in the U.S., both thermal or slow-neutron
reactors, are the Light-Water Reactor (LWR) and the
High Temperature Gas-Cooled Reactor (HTGR). The LWR
uses ordinary water** both as a coolant to transport
the heat released in fission and as a moderator to slow
down the fast neutrons produced in fission. There are
two types of LWR's, the Pressurized Water Reactor (PWR)
and the Boiling Water Reactor (BWR), the distribution
of which in the U. S. is about 60% PWR and 40% BWR. The
HTGR is helium-cooled and graphite-moderated. There is
only one HTGR now operating in the U. S. (Peachbottom,
45 MW) and a second (Fort Saint Vrain, 330 MW) under
construction, but gas-cooled reactors are predominant
in England. (However, the English reactors are cooled
with CO_2 and operated with fuel differing from that of

---

*A fourfold increase in the price of crude uranium would
raise power-plant fuel costs probably less than twofold
for the following reason: $3 invested today by a power
plant in a fuel rod is roughly $1 for crude uranium, $1
for enrichment, and $1 for cladding. Raising the crude
price to $4 from $1 doubles the fuel rod cost from $3 to
$6. But the plant would then be reoptimized to use
slightly less fuel at the new higher price.

**A reactor operating with heavy water (HWR) is used
in Canada.

the U. S. HTGR.)   A brief discussion of these reactor
types, their performance characteristics, and pollutants
and hazards associated with their operation is given
below.

4.3.2. Description of Reactor Types

Pressurized Water Reactors. A PWR and its associated
power plant are illustrated in Figure 4-1. The fuel,
UO_2 enriched to 2.5-3.2% U^{235}, is sealed hermetically
in zirconium-alloy tubing, which constitutes the first
barrier against the escape of highly radioactive fission
products. Assemblies of the fuel rods thus formed are
surrounded,in a steel pressure vessel, by flowing water
entering at 545 F, leaving at 610 F, and held at 2250
psi to prevent boiling. The water slows down neutrons
produced in fission and increases their probability of
reacting with U^{235} to such an extent that the fuel constitutes a critical mass capable of sustaining the chain
reaction. The reaction is held at steady state by use
of a variable amount of boron, partly as movable control
rods and partly as boric acid dissolved in the water. A
pump circulates the water through the reactor, past a
steam-cushioned pressurizer which maintains constant
pressure, and through the steam generator where the primary water is heat exchanged with secondary, boiling

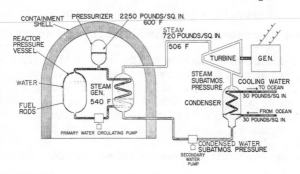

Figure 4-1. Pressurized Water Nuclear Power Plant
(Benedict, 1971)

water to produce steam at 506 F and 720 psi. The steam
is expanded in a turbine driving an electric generator,
and is then condensed at subatmospheric pressure by heat
exchange with cooling water, with the condensate being
recycled by pumping.

Boiling Water Reactor. The BWR is similar to the PWR
except that the primary water, which enters the reactor
at 376 F and leaves at 546 F, is held at a lower pressure,
around 1000 psi, and allowed to boil in the reactor.
Steam and water flowing past the fuel are separated; the
water is recirculated and the steam flows directly to the
turbine after which it is condensed and returned to the
reactor. The BWR needs no separate steam generator.

High-Temperature Gas-Cooled Reactor. The HTGR uses
highly enriched homogeneous fuel consisting of dicarbides
of uranium and thorium dispersed in a graphite matrix.
The weight distribution among the heavy metals in the
initial fuel is 5.03% U^{235}, 94.6% Th^{232}, and 0.37% U^{238}.
The initial fissile material is U^{235}, and Th^{232} serves
as the fertile material by producing fissionable U^{233}.
The major portion of the energy production therefore
results from the relatively inexpensive thorium. The
fuel elements are clad with graphite in order to permit
high temperature and therefore high thermal efficiency.

Helium enters the reactor at 760 F, leaves at 1430 F
and 700 psia, flows through a steam generator where steam
at 2500 psia and 1000 F is produced, and then returns
through a compressor to the reactor. The steam passes
through a turbine for power production.

4.3.3. Performance Characteristics

PWR and BWR thermal efficiencies are 32-33%. Current BWR
designs call for specific powers of about 22 kW(t)/kg U;
performance is predicted to improve to 30 kW(t)/kg U in
the future. At 3% enrichment these values correspond to

about 660 and 900 kW(t)/metric ton fuel (UO_2). The
current-design and predicted performances of the PWR
are 35 and 46 kW(t)/kg U. Fuel burnup specifications for
light-water reactors are in the range 30,000-35,000
MW(t)-days/metric ton U. The spent fuel has an enrichment of about 1-1.5% in residual U^{235} plus Pu^{239} made
from U^{238}.

The HTGR has a thermal efficiency of 39.2%, a specific
power of 1,100 kW(t)/kg, and a burnup of 100,000 MW(t)-
days/ton(U+Th). This performance is extremely high compared with light-water reactors.

The cost of electricity from light-water reactors
varies among different plants. Table 4-2 illustrates
the economics of a PWR power plant having two 940 MW
generating units and expected to be operational in 1974-
75 near Fredericksburg, Virginia.

Table 4-2. Cost of Electricity from Pressurized-Water
Nuclear Reactor (Benedict, 1971)

Item	Value
Unit Investment Cost of Plant, C	$255/kW
Annual Capital Charge Rate, i	0.13
Kilowatt-Hours Generated per Year per Kilowatt Capacity, k	5256 hr
Heat Rate, h	10,400 Btu/kWh
Cost of Heat from Fuel, f	18¢/10^6 Btu
Cost of Electricity, mills: Plant Investment, 1000 Ci/k Operation and Maintenance Fuel, 10 hf Total	6.31 mills/kWh 0.38 " " 1.87 " " 8.56 mills/kWh

The plant cost, fuel cost, and electricity cost are
$255/kW, 18¢/10^6 Btu, and 8.56 mills/kWh. A coal-fired
steam plant of the same output, at the same location,
and without sulfur dioxide control, costs $202/kW and
produces electricity at the same cost as the nuclear
plant if the coal cost is 36.2¢/10^6 Btu. A present-day
characteristic of light-water nuclear plants is the higher capital cost and lower fuel cost compared with fossil
fuel plants.

4.3.4. Pollutants and Hazards

Introduction. The waste heat, radioactive emissions,
and radioactive wastes of light-water nuclear reactors
constitute potential environmental problems, and reactor
safety continues to receive much public attention. Some
characteristics of these problems and techniques for
dealing with them are described below.

Waste Heat. The disposal of waste heat from nuclear
plants is discussed in Section 2.3. A brief summary is
given here. Waste heat is carried from the plant in the
cooling water flowing through the condenser. The temperature rise of the water is typically 20 F, and the
flow rate for a 1000 MW, 33% efficient plant is about
1500 cu. ft./sec. The cooling water is drawn from either
the ocean or some other natural source and is discharged
after one pass through the condenser, or it is recirculated through cooling towers or cooling ponds. Thus the
waste heat is discharged into the biosphere either (a)
directly as 130 million cu.ft./day of water initially
at 20 F above the source temperature, (b) as warm dry
air from a dry cooling tower, or (c) as warm wet air
from a cooling pond or wet cooling tower, simultaneously
evaporating 1 to 2% of the cooling water (more in summer).
Disposal of the heat contained in the warm water without
adverse environmental effects is a more serious problem

for light-water nuclear plants than for fossil-fuel
plants. The nuclear plants, which have a lower thermal
efficiency and do not discharge part of their waste heat
directly into the atmosphere in stack gases, discharge
about 2/3 more heat to the cooling water. The control
of thermal pollution from light-water nuclear plants by
use of cooling ponds, wet cooling towers and dry cooling
towers increases the cost of electricity, relative to
that from a plant with once-through cooling, by about
0.08, 0.1-0.2, and 0.9-1.2 mills/kWh (Section 2.3).
Radioactive Emissions from Power Plants. Escape of the
enormous amounts of radioactivity contained in nuclear
power plants is held to insignificant levels by multiple
barriers. The kind of barriers used depends upon the
type of reactor. The following discussion of precautions
taken in the operation of pressurized water reactors is
illustrative.

Fission products which collect in the UO_2 fuel are the
main source of radioactivity. Whenever the zirconium
tubes leak, as they sometimes do, a small fraction of
those fission products which are water-soluble appear in
the primary water. Some of the fission products, such
as refractory oxides, are water-insoluble and therefore
not troublesome in the reactor. The primary water also
carries corrosion products made radioactive by neutron
activation. Continuous purification by filtration and
ion exchange is used to control radioactive content of
the primary water, and the escape of radioactivity from
the primary water is effectively controlled by using a
leak-tight pressure vessel and piping system for primary-
water circulation. A leak-tight steel and concrete containment shell housing the primary system provides a
further barrier against the escape of radioactivity in
the unlikely event of a major escape of fission products
from the zirconium tubes and then from the reactor. The

secondary-water system is designed to be leak-tight in
order to prevent the further escape of any radioactivity
escaping, in the steam generator, from the primary water
into the lower-pressure secondary water. No leakage can
occur in the condenser from the secondary water into the
cooling water because the cooling-water pressure exceeds
that of the condensing steam. Stringent precautions are
taken during refueling, when the reactor is shut down and
the pressure vessel is opened in order to concentrate,
package and confine radioactive materials present.

The control of radioactivity from boiling water reactors is somewhat more difficult than the procedures
described above because the same water that passes over
the fuel elements also passes through the condenser,
where it is separated by a metal wall from the cooling
water. A rupture of a condenser tube could allow some
of the slightly contaminated primary coolant to mix
with the secondary coolant. Impurities such as certain
metal oxides have a long enough half-life in the activated state to make such contamination possible.

Release of radioactivity to the environment from
light-water reactors can be controlled to any degree
deemed necessary, but with increasing cost. All U. S.
nuclear power plants are monitored by the U. S. Public
Health Service and have been found to add to the environment only a minute fraction of the amount of
radioactivity naturally present (Public Health Service,
1970). As an example, the concentrations of radionuclides in the water discharged from a 1000 MW power
plant employing a pressurized water reactor are given,
along with the half-lives, in Table 4-3. Even the
highest concentration, that of tritium, is far below
the maximum permissible concentration. Radioactive
release in the gaseous effluents of the same plant is
5 to 200 curies/yr, which is mainly noble gases (Kr[85],

Table 4-3. Radioactivity in Water Discharged from a
1000 MW Pressurized Water Nuclear Reactor (SCEP, 1970)

Isotope*	Half-Life, Years	Concentration Microcuries**	Ratio of Concentration to Maximum Permissible Concentration for Unrestricted Water Bodies
H^3	12.3	3.8×10^{-6}	7×10^{-4}
Cs^{134}	2.3	2.1×10^{-12}	2×10^{-7}
Cs^{137}	27	3.4×10^{-12}	2×10^{-7}

* Ten other isotopes reported at concentrations less than
2×10^{-10} times that permitted, and four at less than 3×10^{-8}
times that permitted, are not included here.

**One curie is the amount of radionuclides sufficient to
produce 3.7×10^{10} disintegrations/sec, without regard to
the energy associated with the emission.

half-life 10.8 years; some Xe^{133}), coolant activation
products, and traces of halogens*** and particulates.
This rate of release is not regarded as a problem.

Radioactive Wastes.   Spent fuel assemblies removed from
light-water reactors are allowed to cool for about 4
months before reprocessing. The cooled assemblies are
sealed in shielded containers designed to withstand shipping accidents, and then transported to a reprocessing
plant where they are cut open, their contents are dissolved in acid, and uranium and plutonium are recovered
by solvent extraction. The radioactive fission products
are concentrated and stored in solution in double-walled
containers for about 5 years, at which time present
Atomic Energy Commission regulations require that the
solution be evaporated to dryness and that the solid
fission products be sealed in stainless steel containers.
Finally the containers are shipped to a national

***The four pressurized water reactors operating at
greater than 100 MW in 1968 reported negligible effluent
concentration in this category.

radioactive waste repository for long-term storage.

The technology for shipping and reprocessing radioactive fuel has been developed and proved safe for
present purposes by several years of operation, but
additional precautionary measures will be required in
the future if nuclear power production is to grow as
expected. With present reprocessing techniques, the
amount of radioactive emission occurring during reprocessing is about 100 times larger than that occurring at
the power plant, and the long-lived radionuclides Kr^{85}
and H^3 are released in their entirety. Measures for
retention of these species during reprocessing will be
required when the amount of fuel is greater than now.
The operation of a 1000 MW light-water reactor is associated with reprocessing 96 tons of U/yr, which
produces 9600 gallons/yr of high-level liquid waste
requiring storage. This waste is converted to 96 cu.
ft./yr of solids. It is clear that the volumes of
radioactive materials to be handled in the future will
be large.

Transportation of radioactive cargoes, which is regulated by the Atomic Energy Commission and the Department
of Transportation, must be accomplished without effects
adverse to the environment or to public health and
safety, and the chance of a black market in nuclear
materials must be avoided. The cargoes include fuel,
fuel elements, spent fuel, and fuel wastes travelling
mainly by rail in steel or lead casks to and from
nuclear reactors, processing and reprocessing plants,
gaseous diffusion plants, and waste-disposal sites. An
increasing number of shipments will contain more plutonium (which is very hazardous to ship), than at present,
since present light-water reactors will start to convert
to recycled plutonium as fuel in 1973. Shipments as
large as 3 tons of plutonium in a 100-ton cask are

expected (Nussbaumer, 1969). The number of casks of spent
fuel shipped per year is expected to rise from 30 in 1970
to about 9500 in the year 2000. Assuming that the average
trip length and the Department of Transportation's overall
rail-accident rate continue to be about 500 miles and
about 0.3 serious accidents per million rail miles, the
huge casks of spent fuel would be involved in about 1.4
serious rail accidents per year in the year 2000. Since
it is not feasible to eliminate rail accidents completely,
an acceptable balance must be achieved, in the transportation of radioactive cargoes, between the extent of
adverse effects and the amount of spending allowed for
precautionary measures.

The present plan for disposal of high-level radioactive wastes calls for burial 1500 ft. underground in
a Central Kansas bedded salt mine. This proposal was
recently examined, and approved, by the Committee on
Radioactive Waste Management of the National Academy of
Sciences-National Research Council (NAS-NRC, 1970), but
it remains the subject of considerable debate. Liquid
high-level wastes are now stored in concrete encased
stainless steel vaults in Idaho, South Carolina, and
Washington, and there is general agreement that a better
method is needed. Burial in salt beds has the following
advantages: a suitable salt bed effectively separates
radioactive waste and its radiation from the environment
for 1000 years or longer and reduces the probability of
release by accident or malicious acts; bedded salt
exhibits plastic flow which relieves stress concentrations produced by mining or heat from the wastes, heals
fractures in the salt and, at the temperature of the
wastes, seals the containers in cells of crystalline
salt; salt beds permit more rapid heat dissipation than
do other types of rock; loss of the salt resources used
in this way is negligible. The Kansas salt bed was

chosen because of its favorable geological properties,
including low probability of earthquakes, a history of
250 million years of dryness, simple structure, low
relief, and tectonic stability.

The NAS-NRC committee generally approved a proposed
AEC demonstration project in Lyons, Kansas, provided a
series of research and testing are carried out, some
prior to operation of the project, including: determination of bed thickness and uniformity by core drilling
in order to plan the location of shafts and distribution
of rooms; a survey of neighboring gas and oil wells to
determine whether they are adequately plugged to prevent
ground water from entering the salt; a study of the
best mode of mining and backfilling to avoid future
subsidence; laboratory determination of the temperature
rise to be expected in the salt due to gamma radiation;
and securing the zone around the site to prevent accidental drilling which might affect the integrity of the
site. Studies which need not be completed before work
begins, but which could expose reasons for discontinuing
the project, are in three categories: exploration for
and mapping of aquifers around the salt beds; further
investigations into the thermal and mechanical properties
of the bedded salt and its associated stratigraphic units;
investigation of means of removing salt mined from the
beds. The panel also cited a number of lower-priority,
longer-term needs including: development of a contingency plan for retrieving the wastes, should it become
necessary; development of a monitoring system to check
water quality and temperature, solid temperature, radiation levels, seismic activity, surface subsidence,
plugged wells, and changes in the surface biota after
sealing the mine; determination of whether subsurface
leaching poses any threat to the future integrity of
the disposal; and determination of the extent to which

erosion might affect the site over the next 500,000 to
1,000,000 years. The Kansas State Geological Survey,
taking issue with the AEC and the NAS-NRC report, maintain that a number of questions should be resolved
before proceeding with the demonstration. Of particular
concern is the possible migration of waste containers
through the salt, the possibility that thermal expansion
of the salt might crack overlying rock layers and allow
ground water to seep in, and possible unforeseen radioactive interactions. The fact that no retrieval plan
has been developed receives strong criticism from Kansas.

At present there is no alternative to the salt mine
disposal method that is thought to be safer. The enormous
responsibility of safeguarding hazardous material for a
thousand years requires that reasonable criticism of the
method be given careful consideration, and that exploitation of the method should proceed gradually, with continued evaluation.

Reactor Safety. There are at present no identified features of LWR's which make them inherently unsafe. Acute
low-probability accidents which have the potential for
affecting public safety involve a sequence of events
including initiation, the rapid release of large amounts
of thermal energy, conversion of energy into mechanical
work, and breach of containment coincident with gross
release of the core fission product inventory. The main
factor which must be taken into account in avoiding such
a sequence is that LWR's contain a reservoir of pressurized coolant which, if release is initiated, is capable
both of generating missiles and gas overpressures which
may threaten containment integrity and of flashing to
vapor, thereby depriving the core of heat removal after
shutdown. The potential for damage during an accident
is increased by the metal-water reaction involving
molten cladding.

A safety issue which recently received much public
attention concerns the adequacy of the emergency core-
cooling system now used on light-water reactors. In
experiments at the National Reactor Testing Station in
Idaho, the safety response to a deliberately induced
failure of a simulated reactor was inadequate. Since
no actual reactors were involved in the tests there is
some question about the applicability of the results
to real systems, but there is no doubt that such an
important issue must be resolved.

The cost of safety features added primarily to protect the general public over and above those added to
protect the large economic investment is not always
clearly identifiable, but the largest single item by
far in this category is the containment building of the
typical PWR. According to estimates by Oak Ridge
National Laboratory, containment requirements add about
6% to the overall busbar power cost. This figure is an
indication of the high emphasis placed on safety in the
U. S. In the Soviet Union, for example, it is standard
practice to avoid this cost by using confinement buildings similar to the typical steam-plant building in the
U. S.

4.3.5. Future of Present Reactor Types
There is little doubt that nuclear fission will eventually become as important as fossil fuel combustion
as an energy source for electric power generation. The
required nuclear power will have to come from presently
available reactor types for some time to come, after
which time breeder reactors will be available as competitors. The future of present reactor types is
therefore strongly influenced by breeder development,
for which reason further discussion of this topic is
postponed until after breeders are discussed.

4.4. Breeder Reactor Technology

4.4.1. Introduction

Breeder reactors derive their name from the production
of a net amount of fissionable material. Breeding is
achieved by reacting, for the primary purpose of power
production, a fuel containing both fissionable and
fertile materials under conditions in which a sufficient
number of neutrons is released by fission to supply losses, to produce, by conversion of fertile material, more
fissionable material than is consumed, and to propagate
the fission reaction. The two most familiar breeding
cycles, $U^{238} \rightarrow Pu^{239}$ and $Th^{232} \rightarrow U^{233}$, are illustrated
below:

Uranium Breeding Cycle

$$_{92}U^{238} + _{0}n' \rightarrow _{92}U^{239}$$

$$_{92}U^{239} \rightarrow _{93}Np^{239} + \beta^{-} \text{(half-life: } 24 \text{ min.)}$$

$$_{93}Np^{239} \rightarrow _{94}Pu^{239} + \beta^{-} \text{(half-life: } 2.3 \text{ days)}$$

$$_{94}Pu^{239} + _{0}n' \rightarrow \text{Fission Products} + x_{0}n' \quad (x>2)$$

Thorium Breeding Cycle

$$_{90}Th^{232} + _{0}n' \rightarrow _{90}Th^{233}$$

$$_{90}Th^{233} \rightarrow _{91}Pa^{233} + \beta^{-} \text{(half-life: } 22 \text{ min.)}$$

$$_{91}Pa^{233} \rightarrow _{92}U^{233} + \beta^{-} \text{(half-life: } 27 \text{ days)}$$

$$_{92}U^{233} + _{0}n' \rightarrow \text{Fission Products} + y_{0}n' \quad (y>2)$$

The uranium cycle, which uses fast (highly energetic) neutrons, is carried out in fast breeder reactors which use coolants (e.g., liquid sodium or pressurized helium) that do not slow neutrons down. In the cycle an atom of fertile U^{238} absorbs a neutron and emits a beta particle to become neptunium, which then undergoes beta decay to become fissionable Pu^{239}. Fission of a Pu^{239} atom with a neutron produces fission products, releases energy, and produces about 2.6 neutrons. Loss of neutrons to nonproductive ends is small in this cycle, and after one neutron goes to continue the fission chain reaction about 1.5 neutrons are left to react with U^{238} to make 1.5 atoms of Pu^{239}, one of which replaces the atom used in fission and the other 0.5 is the net plutonium gain. This cycle permits nearly all the U^{238} in natural uranium to be used as fuel. The thorium cycle is similar to the uranium cycle except that it works best in a thermal breeder reactor, where it uses thermal (relatively slow) neutrons. The fertile isotope of thorium Th^{232} is converted to protactinium which, by beta decay, becomes fissionable U^{233}.

Three types of breeder reactors being studied in the U. S. are the Liquid Metal Fast Breeder Reactor (LMFBR), the Gas-Cooled Fast Breeder Reactor (GCFR), and the Molten Salt Breeder Reactor (MSBR). The LMFBR and GCFR use the uranium cycle while the MSBR uses the thorium cycle.* These reactors are discussed below.

----

*Another reactor concept based on the thorium cycle is the Light Water Breeder Reactor (LWBR) under development in the Shippingport program of the AEC Division of Naval Reactors. This reactor is not assessed here because published cost evaluations are not available. There is need for design studies from industry to determine the economics of the LWBR.

4.4.2. Reactor Types and Expected Performances

Liquid Metal Fast Breeder Reactor. The LMFBR , shown
diagrammatically in Figure 4-2, uses fuel consisting of
approximately 80% (by weight) UO_2 and 15% PuO_2 in small
diameter (0.25") stainless-steel clad tubes operating
at temperatures up to 1250-1300 F. Fuel pins are mounted in clusters of 127 to 217 pins forming fuel assemblies which are the basic building blocks of the reactor
core. Surrounding the core both axially and radially
is a blanket region made up of pins containing natural
or depleted uranium. The fuel assemblies are immersed
in liquid sodium coolant which flows at low pressure
through the reactor, entering at 750 and leaving at
1150 F. Since the primary sodium becomes intensely
radioactive in the reactor, a secondary, nonradioactive
sodium coolant loop is interposed between the primary
sodium loop and the steam generator. Two different
designs of the primary system are being employed--the
pot type system in which all components in the primary
circuit are submerged in a sodium-filled pot and the
loop type system (Figure 4-2) in which the reactor
vessel, pump, and heat exchanger are linked by piping
through which sodium is circulated.

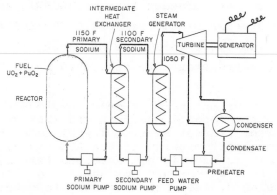

Figure 4-2. Liquid Metal Fast Breeder Reactor (Benedict, 1971)

Demonstration LMFBR plant designs are based on specific powers of about 1000 kW(t)/kg fissile [which at
an enrichment of approximately 15% fissile corresponds
to 150-210 kW(t)/kg(U+Pu)] and fuel burnups of 100,000
MW(t)-days/metric ton (U+Pu). The LMFBR is expected to
have a doubling time* of 8-9 years, a thermal efficiency of about 40%, and a breeding ratio of about
1.5**.

Few projections of capital cost for LMFBR power
plants have been made, but there is hope that commercial
designs will cost no more than $25 to $50/kW more than
light-water-reactor power plants. (At 80% load factor,
$100 would just offset the entire fuel cost of the LWR;
the LMFBR would have to save around 1 mill/kWh on the
fuel cycle in order to offset a $50/kW increase in
capital cost.)   A recent study (Edison Electric
Institute, 1970) indicates that the capital cost for
a 1000 MW commercial LMFBR (mid-1980 startup) will be
approximately $240/kW. Other studies had placed the
capital cost in the range of $240-$270/kW. The same
EEI study indicates that LWR capital costs will be
approximately $175/kW in 1985, which assumes that
plant improvements between now and 1985 will lower
LWR from the EEI estimate of $220/kW for 1975 startup.

Gas-Cooled Fast Breeder Reactor.   The GCFR is cooled
with helium at 1250 psi and therefore benefits from
technology already developed for the HTGR. Like the
LMFBR, the fuel used is UO_2-PuO_2. The GCFR power
plant, shown schematically in Figure 4-3, includes
----
*Time required to double the inventory of fissionable
material.

**Mass ratio of fissionable material produced to that
consumed.

the reactor core, steam generators, and helium circulators integrated within a prestressed concrete
reactor vessel. Cavities for the steam generators
and helium circulators are separated from the core
cavity to facilitate repair and maintenance. The
circulating helium flows down through the reactor
core, through the steam generators, and then through
compressors back to the top. Unlike sodium, helium
does not become radioactive under neutron bombardment; so steam can be generated directly from
primary helium without need for a secondary coolant
loop. The helium enters the core region at 600 F
and leaves at 1200 F. The core, which consists of
approximately 280 fuel elements each containing 250
stainless-steel-clad fuel rods, 30 control rods,
and 200 radial blanket elements, is suspended from
a grid plate which is cooled along with the control-
rod drives by helium returning from below. Fuel is
loaded and unloaded through the bottom of the core.

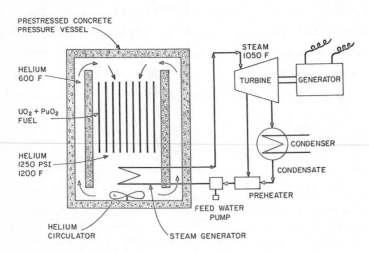

Figure 4-3. Gas-Cooled Fast Breeder Reactor (Benedict,
1971)

The helium circulation system uses six single-stage,
axial-flow helium compressors, driven by integrally
mounted steam turbines, which provide a compression
ratio of about 1.05. The steam turbine is a single-
stage machine which operates in series with the main
power turbine; the six units, operating together, carry
the full main steam flow. Steam turbine drive permits
short-term overspeeding to ensure adequate cooling in
the event of loss of coolant pressure, even with a reduced number of circulators in operation.

The GCFR is expected to have a thermal efficiency of
about 40%, a doubling time of 8-9 years, and a breeding
ratio of about 1.5. The use of carbide fuel in advanced
plants is expected to increase the breeding ratio to
1.6 and to decrease the doubling time to approximately
5 years. The design burnup and specific power are
100,000 MW(t)-days/metric ton(U+Pu) and 800-900 MW(t)/kg
fissile.

Molten Salt Breeder Reactor. The MSBR uses a fused salt
mixture of uranium and thorium fluorides dissolved in
beryllium and lithium fluorides for both fuel and coolant.
The moderator is unclad graphite, which can be used bare
since the salt does not wet graphite and will not penetrate into its pores if the pore size of the material is
less than about 1 micron. The salt mixture melts at
813 F, which requires that all parts of the system containing salt be heated. The vapor pressure is less than
0.1 torr at 1300 F, the viscosity at reactor operating
conditions is similar to that of kerosene, and the volumetric heat capacity is about the same as that of
room-temperature water. The reactor vessel and all
parts of the system that contact the salt are made of
Hastelloy N, a material especially developed for use
with molten fluorides.

The salt circulates through the reactor, where the

graphite moderators serve also as a separation medium
for two fuel regions, then through the circulation
pump, and then through a heat exchanger where heat is
transferred to a coolant salt which circulates through
a steam generator. The coolant salt is similar to the
fuel salt but contains no uranium or thorium. A bleed-
off stream of fuel salt is continuously withdrawn and
processed, in an adjacent chemical plant, to remove
impurities such as rare-earth fission products, to
segregate protactinium until it decays to U^{233}, and to
recover the U^{233} made by breeding. The purified salt
is recycled.

The breeding ratio is expected to be 1.06-1.07.
The doubling time may be as high as 20 years, although
about 10 years is thought to be possible.

4.4.3. Status of Development

Liquid Metal Fast Breeder Reactor. Compared with the
GCFR and the MSBR concepts, far more experience and
progress exists in construction and operation of LMFBR
demonstration units and small power plants. A list of
LMFBR power plants appears in Table 4-4. In June 1971
the President proposed to Congress that an additional
$50 million be authorized to increase, to $130 million,
the U. S. Government contribution to the construction
of the first U. S. fast breeder demonstration reactor
(which would undoubtedly be a LMFBR). (The estimated
total requirement is $300-600 million.) If approved
by Congress, this strong evidence of federal commitment
should ensure start of plant construction in time to
meet 1978 criticality.

Gas-Cooled Fast Breeder Reactor. The GCFR draws on the
experience gained in construction and operation of the
thermal-spectrum high-temperature gas reactor type
presently competing with light water reactors.

Table 4-4. Liquid Metal Fast Breeder Power Plants
(Benedict, 1971)

Country	Reactor	Electric Megawatts	Years Operated
U. S.	EBR-I	0.3	1951-1963
U. K.	DOUNREAY	60	1963-
U. S.	EBR-II	20	1965-
U. S.	FERMI-I	70	1965-
France	RAPSODIE	20	1967-
USSR	BR-60	60	1970-

Under Construction			Scheduled to Operate
USSR	BN-350	150	1971
U. K.	PFR	250	1972
France	PHENIX	250	1973
USSR	BN-600	600	1973-1975

Consequently there exists no specific experimental or
test reactors based on the GCFR concept, but the Peachbottom and Fort Saint Vrain HTGR plants do provide
useful reservoirs of experience with regard to the
primary circuit of the GCFR concept. There is no
present U. S. commitment to construct a GCFR experimental reactor or demonstration reactor.
Molten Salt Breeder Reactor. The projects which contribute to MSBR technology include the Aircraft Reactor
Experiment, a molten salt reactor operated at ORNL in
1954 as part of the Aircraft Nuclear Propulsion Program,
and the Molten Salt Reactor Experiment (MSRE) which was
constructed in 1962-1965, operated on U^{233} in 1965-1968,
and operated on U^{238} in 1968-1969. Although the long
operation of the MSRE proved the technical feasibility
of a molten salt reactor, there is no present commitment to construct a molten salt demonstration reactor.
According to a recent news item (Chem. and Eng. News,

1971) Oak Ridge National Laboratory has let a $850,000
subcontract to a six-company group headed by Ebasco
Services, Inc., for an independent industrial assessment
of the technological prospects and economic incentives
for developing the MSBR.

4.4.4. Research Needs
Introduction.   The potential of the breeder reactor as
a future energy source is unquestionably large and breeders must be developed, but there is some question about
the urgency of the development. The competitors which a
new breeder must face--highly developed light-water reactors and clean combustion processes--must eventually
be replaced since fuel reserves are finite, but the
reserves are not yet scarce enough to permit confident
projection of the time at which breeders because of
decreasing reserves must be brought to commercial real-
ity. Furthermore, it is not yet established whether
the increased capital cost of breeders will be offset
by the decreased operating cost resulting from savings
on the fuel cycle. There can be little argument about
the value of making a wise selection of breeder concept
to be pushed but, since there is uncertainty about the
proper time scale for breeder development, there is
room for debate as to when the choice should be made.
Either a premature or an overly prolonged decision
would be costly both in time and money.

   A comprehensive discussion of research needs in all
the areas related to nuclear power production is clearly beyond the scope of this study and the competence of
its staff. Discussions with professional associates,
however, disclosed a number of unsolved problems which
it appeared appropriate to formulate and include in
this study. In the following pages a brief discussion
of conflicting views about the relative merits of

various breeders is followed by an outline of major
problems to be solved for each type of reactor.
Choice of Breeder Concept. The GCFR and the MSBR both,
on paper, offer potential advantages with respect to the
LMFBR system. In particular, the GCFR system appears
to offer a better breeding ratio and lower doubling time
while the MSBR provides a potential for utilizing thorium
resources in a system with acceptable cost and safety
characteristics. The AEC provides modest funding for
continued development activities of both GCFR and MSBR
systems ($5 million/year to each) and directs most of
the available reactor development budget toward the
LMFBR program (over $100 million/year). This emphasis
is the direct result of AEC evaluations of these breeder concepts conducted up to 1969. The LMFBR has also
been chosen for development by USSR, U.K., and France.

The central issue in the reactor development program
relative to introducing any concept is the ability to
take a feasible system and gather sufficient industrial
and utility support to project it to the demonstration
and commercial plant stage. Such an effort requires
large financial and technical resources. The AEC view-
point is that these resource requirements are such that
the U. S. can afford to fund effectively only one concept at this time. The AEC program calls for the exploitation of that one concept, the LMFBR, by initial
construction of two demonstration plants (starts spaced
about 2 years apart) to insure that there will be more
than one commercial manufacturer of LMFBR plants.
However, funds are being made available to the AEC for
only one plant, which supports the AEC contention that
the funds available are not sufficient to permit emphasis on more than one concept at this time.

The other side of the argument is that an over-
concentration of U. S. effort on one concept leaves

the U. S. vulnerable in the event of technical design and
safety difficulties with the LMFBR concept. Supporting
this position is the U. S. record in the reactor development area, which is replete with examples of failures to
succeed in the development of promising reactor concepts.
The ultimate question is how the prospects for success
of the LMFBR--including the short-term freedom from
detrimental technical problems and the ultimate potential
offered by construction of the LMFBR--compare with the
projected ultimate potential of the GCFR and MSBR.
Liquid Metal Fast Breeder Reactor. One of the main
problems in LMFBR development is the production of an
adequate fuel element. Because LMFBR plants will involve
a higher capital investment than present nuclear plants,
the residence time of the fuel elements between reprocessings must be increased markedly over present guarantee.
Up to now there has been no experience with the need for
changing fuel elements. About two years ago the first
test fuel elements were taken from a British reactor.
The stainless steel sheath was swelled and contained
microscopic voids which apparently are created by neutron
displacement of the Fe atoms, and which collapse upon
annealing. Since the LMFBR fuel elements must withstand
much more intense neutron radiation than that occurring
in present reactors, the metallurgical problems with the
LMFBR elements may be severe. The elements for proposed
LMFBR demonstration plants are designed for a life-time
between reprocessings about 75% as long as that needed
in commercial plants. Development of new fuel elements
is currently under study in the Experimental Breeder
Reactor II, which study will be complemented by work on
the higher-capacity Fast Flux Test Facility.

One of the most important safety issues in LMFBR
development is that of core stability. Present
commercial light water reactors compensate automatically

for any accidental increase in power output. If a section in the core overheats, the water density there decreases, thus reducing the ability of the water to moderate the fast neutrons. Since the probability of U^{235} fission is higher with thermal neutrons than with fast neutrons, the rate of fission decreases and the reactor returns to normal. In contrast to this behavior, an increase in local reactivity in the LMFBR reduces the local sodium density and, through a number of linked physical effects, may increase the number of fast neutrons available for fission of Pu^{239}, which increases the reaction rate still further. Other effects may provide a negative temperature coefficient for reactor stability. The expansion of metal fuels upon heating allows more neutrons to escape and thereby slows the reaction. However, the oxides to be used for LMFBR fuel do not expand as reproducibly as metals, and this effect is of questionable value. Stability may result from the Doppler effect, which is the increase, with increasing temperature, of the neutron-absorption cross section of U^{238}. Since the conversion of U^{238} to Pu^{239} involves a decay process with a half-life of 2.3 days, neutrons absorbed by U^{238} are effectively eliminated in excursions, which take place in a fraction of a second. The magnitude of the Doppler effect in fast reactors is not well established experimentally, but recent work at the Southwest Experimental Fast Oxide Reactor indicates it may be very near the theoretical value and twice as great as is considered adequate for providing stability in small excursions. It is, however, probably too small to prevent a large excursion in the case of a more serious situation such as the existence of a large sodium void in the center of the core, but it could mitigate the consequences of a large excursion. If such a situation could develop, for example by

blockage of local coolant flow, a key question is whether
the additional energy generated in the region of the void
is adequate to propagate void formation. A core design
must be developed which prevents voids and other possible
malfunctions, such as fuel pin failure by melting on
overpower transients, from spreading throughout the core.

Transportation and reprocessing LMFBR fuel will require
more precautions than are necessary for the fuel of present reactors because, in LMFBR operation, the specific
power will be higher [150-210 kW(t)/kg(U+Pu) for LMFBR
versus 34-46 kW(t)/kg U for LWR], the total irradiation
of the fuel to be reprocessed will be higher [fuel burnup is 100,000 MW(t)-day/metric ton (U+Pu) for LMFBR versus
30,000-35,000 MW(t)-days/metric ton U for LWR], and the
cooling time allowed for spent fuel will be smaller (about
30 days for LMFBR versus at least 120 days for LWR). The
choice of a shorter cooling time for LMFBR spent fuel
results from the fact that the higher fissile enrichment
[15% for LMFBR versus 2-3% (feed) and 1-1.5% (discharge)
for LWR] will result in daily carrying charges on capital
investment during the cooling period 10-15 times greater
than those for LWR operation. These three factors will
result in spent fuel which contains much more radioactivity at the time of reprocessing than is the case
for present reactors. Consequently, handling of LMFBR
spent fuel will require better thermal cooling during
shipping and greater safeguards against shipping accidents, and the first operations in the reprocessing
plant will bear a much greater radioactivity and thermal
heat load. Radiation damage of the solvent extractants
may cause some difficulties, but the most serious
problems at the reprocessing plant will be feed preparation, which includes the mechanical and chemical
disassembly and dissolution of the fuel for solvent
extraction separation, and handling radioactive iodine

and krypton gases. Furthermore, the higher fissile enrichment of the discharge core fuel from the LMFBR will
impose more stringent requirements for criticality limitation in processing than is now the case for LWR fuel.
The higher unit-fuel costs which will result from measures taken to deal with these factors are expected to
be offset if the higher fuel burnup can be realized.

There are also some LMFBR problems related to the
use of sodium coolant, some of which have been dealt
with to a considerable extent in previous experience
with sodium-cooled reactors. Hot sodium reacts fiercely
with air and water; so it is essential to prevent leakage from the cooling system and to provide an argon
blanket for open sodium surfaces. It is also opaque,
requiring that refueling and other operations be carried out blind. In addition, commercial reactors will
entail the handling of sodium in unprecedented quantities,
requiring progress in the design of new pumps, valves,
heat exchangers, and other equipment.

The advent of the fast breeder would make the nuclear
power economy of the country increasingly based upon
plutonium. This trend will actually start without the
breeder, in 1973, when present light-water reactors
begin their conversion to plutonium. This conversion
requires certain precautions because of plutonium's
relatively high specific alpha radiation activity (compared with that of uranium and thorium) and plutonium's
biochemical behavior. Since the energy of alpha particles is dissipated a short distance from the source,
an alpha emitter within a human body can cause severe
damage in the surrounding tissues. Plutonium in the
body becomes chemically fixed in bones, where it
irradiates the marrow-producing blood cells. While
present reactors produce modest amounts of plutonium
which are removed in the fuel reprocessing plants,

plutonium in the spent fuel from breeders and plutonium-
fueled light-water reactors will be far more concentrated.
For these reasons the development of transportation and
reprocessing procedures capable of handling the risks
associated with increased quantities of plutonium is
very important.

The plutonium to be used in fast breeder reactors,
unlike the enriched uranium used in present reactors,
is relatively easily processed into a form suitable for
use in nuclear weapons. It has been suggested that the
vastly increasing use and shipment of plutonium make
possible the theft, by a hostile power or criminal elements, of enough plutonium (10 lbs would do) to make a
nuclear weapon. In addition, there is the increased
possibility of diversion of plutonium from reactor use
to military use by non-nuclear powers. Such action is
prohibited by the Nuclear Non-Proliferation Treaty to
signatories of that document, but the International
Atomic Energy Agency, which is supposed to enforce the
provisions, has very limited power actually to do so.
Furthermore, a number of nations have not signed the
treaty. This problem needs careful attention.

Gas-Cooled Fast Breeder Reactor. Since the GCFR uses
the same breeding cycle as the LMFBR, the problems discussed above involving the transportation and reprocessing of fuel and the health and theft hazards associated
with plutonium use also apply to the GCFR. However,
there are a number of differences in the problems of the
two reactors some of which arise because of the use of
different coolants. The heat transfer capability of
helium is not as good as that of sodium, but its trans-
parency, chemical inertness, and freedom from bubble
formation avoid some of the problems occurring with
sodium. The GCFR will be dependent upon the technology
of handling large amounts of hot gas at high pressure,

but this problem is being solved through experience with high-temperature gas reactors in the U. S. and Britain.

One of the key technical problems is that of maintaining adequate fuel surface temperatures in the event of a postulated loss-of-coolant or depressurization accident. The reactivity increase due to the loss of coolant would be only a minor problem, but the decrease in coolant mass flow rate upon depressurization would cause a large decrease in core cooling ability. The ability of the design to accommodate this accident depends in large measure on the assumed depressurization rate and operability of heat removal equipment. A second major problem involves the development of an adequate fuel element. In order to exploit the high primary coolant-temperature capability of the GCFR concept, a fuel element cladding capable of operating at surface temperatures considerably in excess of LMFBR practice is required. However, presently to capitalize on the LMFBR fuel development program experience, the GCFR fuel-element cladding temperatures have been derated and a pressure-equalizing venting system incorporated to eliminate differential pressure across the cladding throughout its life. The successful development of the present venting system and the ultimate development of fuel elements to operate at higher limits of the cladding surface temperature represent formidable obstacles to achievement of a reliable fuel system--a requirement which present GCFR assessments assume will be met.

Molten Salt Breeder Reactor. One of the principal questions about the MSBR has been whether the fuel and blanket salts should remain separate. Difficulties in using graphite as a structural material for separating the two salts, and chemical advances which have made it possible to perform the necessary reprocessing

operations with all of the fission products in one fluid
rather than two, provided the impetus for changing the
MSBR concept to one in which the fuel and blanket are in
a single fluid. Significant research and development is
still required on the chemical processing problems encountered in separating U^{233}, Th^{232}, and rare earth
fission products in a single salt stream.

A second major problem in the development of the MSBR
is that of scaling up molten salt pumps and heat exchangers. Although salt has been handled successfully on a
small scale at Oak Ridge National Laboratory, the presence of radioactivity in the salt, the need for pressure
relief against high pressure steam, and the problem of
salt cleanup in the event of tube leakage are some of
the design and maintenance complications. Remote maintenance of the molten-salt fluid-fuel reactor will be
required because of the intense gamma radiation in equipment outside the reactor caused by activation of sodium
and fluorine in the salt, the presence of fission products, and activation of the structural materials by
delayed neutrons in the circulating salt.

It is desired that the fission products be kept at a
low concentration in the core of the reactor. Molten
Salt Research Reactor experience with dilute solutions
of fission products has shown that there is some deposition of the noble metal fission products such as
tellurium, ruthenium, and molybdenum on the surfaces of
Hastelloy N as well as on the surfaces of the graphite.
At the same time, a large fraction of these noble metals
also appear in the gas stream, presumably as metallic
colloids. While the MSRE provided information concerning
fission product behavior in molten-salt reactor systems,
additional information is required relative to fission
product deposition on materials.

One final problem is that of radiation damage

suffered by the Hastelloy N vessel material in molten
salt reactors, specifically the loss of high-temperature
ductility and a reduction in the creep-rupture life due
to the collection of helium at grain boundaries. It is
thought that some progress may be made on this front by
the addition of titanium to the alloy, but this must be
further tested.

4.5. Nuclear Fusion

Though the authors have no special knowledge in the area
of nuclear fusion, they did have discussions with people
who are knowledgeable in the area, read recent papers on
the subject (see, for example, Rose, 1971), and considered fusion in comparison with other energy sources.
Considering the fact that nuclear fission, through the
breeding process described above, offers thousands of
years of energy, fusion research should be treated as an
area of pure science, declaring no motivation from need
for application, and allowed to proceed at an efficient
pace with no element of financial support keyed to
practical need.

References

Atomic Energy Commission, 1970. "Potential Nuclear Power Growth Patterns", Report Wash-1098, December, 1970.

Benedict, M., 1971. "Electric Power from Nuclear Fission", Proceedings of the National Academy of Sciences 68, pp 1923-1930.

Chemical and Engineering News, 1971. August 2, Vol. 49, No. 31, p 27.

Edison Electric Institute, 1970. EEI Reactor Assessment Panel, New York.

NAS-NRC, 1970. "Disposal of Solid Radioactive Wastes in Bedded Salt Deposits", Committee on Radioactive Waste Management, National Academy of Sciences-National Research Council, Washington, D. C., November, 1970.

Nussbaumer, D. A., 1969. Proceedings of the Conference on the Transportation of Radioactive Material, University of Virginia, Charlottesville, 26 to 28 October (Clearinghouse for Scientific and Technical Information, Springfield, Virginia.).

Public Health Service, 1970. "Nuclear Power Reactors and the Population", Report BRH/OCS 70-1, p 14, January 1970.

Rose, D. J., 1971. "Controlled Nuclear Fusion: Status and Outlook", Science 172, No. 3985, pp 797-808.

SCEP, 1970. Man's Impact on the Global Environment-- Assessment and Recommendations for Action, Report of the Study of Critical Environmental Problems, M.I.T. Press, Cambridge, Mass., p 499.

Chapter 5

CENTRAL-STATION POWER FROM FOSSIL FUEL

5.1. Introduction

The first requirement of an electric power generating system is that it perform reliably. In the power area as in
most of man's fields of endeavor, his strengths are his
weaknesses. The premium put on reliability makes a substitution for one of the parts of the system prior to
proof of operability an error in judgment. The resulting
conservatism, not just understandable but laudable, makes
change occur slowly--sometimes too slowly.

It was the initial objective of this study to include
here a presentation of most of the newer ideas relating
to power-plant operation. Limitations on time, however,
have forced a limitation on objectives. This chapter is
in consequence restricted to a consideration of (1) improved gas turbine and mixed heat engine cycles because
one of the combinations looks so promising, (2) magnetohydrodynamic power generation because it has been in the
public eye so long as a revolutionary concept, (3) superconducting electric generators because they offer great
promise, and (4) fuel cells for large power.

5.2. Improved Gas Turbine and Mixed Heat-Engine Cycles

5.2.1. Introduction

Present use of gas turbines for stationary power plants
is confined largely to peaking power applications. Design
philosophies associated with that use tend to reflect the
conservatism of the power industry. Developments in aviation gas turbines, however, make increasingly appropriate a consideration of (1) base-load 200-500 megawatt
turbines for use in 1000 MW stations and of (2) advanced
open-cycle systems using both gas and steam turbines.
Differences between design philosophies for stationary
power plants and for aviation propulsion systems may be
expected to diminish as it becomes increasingly clear

that, whereas the steam turbine has reached its peak of
performance, the gas turbine is still on the upward march.
   This section will present briefly a picture of past
advances in gas turbine system materials, evidence of
present availability of materials properties that permit
confident projection of attainable performance, and comments on turbine blade cooling. Projections of base-load
component performance and of system performance, using
low-Btu clean gas as a fuel, will then be presented, followed by an economic assessment of the system. Finally,
a projection of performance and an economic assessment
will be presented for a combined gas turbine-steam turbine system operating on clean low-Btu gas.*

5.2.2. Projections of Base-load Technology
The gas-turbine cycle or Brayton cycle is based on the
almost-adiabatic compression of air, the almost constant-
pressure combustion of fuel with that air, and the
almost-adiabatic expansion of the hot gases back to atmospheric pressure. In the simplest arrangement the
compressor and turbine are on the same shaft, and the
difference between the work done by the turbine and
that required by the compressor is the net work output.
Improvements in efficiency come from increasing the turbine inlet temperature and the compressor pressure ratio,
and raising the compression and expansion efficiencies of
the compressor and turbine. If the turbine exhaust gas
is to serve as a heat source for steam generation, its
expansion ratio need not be so great, and the compressor
pressure ratio is correspondingly reduced. Consequently,
----
*This section is based largely on a comprehensive report,
"Technological and Economic Feasibility of Advanced Power
Cycles and Methods of Producing Nonpolluting Fuels for
Utility Power Stations," by F. L. Robson, A. J. Giramonti,
G. P. Lewis, and G. Gruber; United Aircraft Research Lab-
oratories to National Air Pollution Control Administration,
December, 1970.

for combined gas-steam cycles high inlet turbine temperature is more important than high compression ratio; for
the simple cycle, however, a high compression ratio is a
prerequisite for high efficiency.

Materials improvements have permitted turbine inlet
temperatures to rise about 20 F per year since 1959,
starting at 1400 F. Added to this is a rise due to improved cooling techniques which makes inlet turbine temperatures of 2000 F feasible now (see Figure 5-1a). A
measure of the permitted rise in inlet turbine temperature is the linear march, with time, of the temperature
to produce, in stressed material, one percent creep in
1000 hours (a) at 20,000 psi for blade alloys (1200 F
in 1946; 1800 F in 1966) or (b) at 10,000 psi for vane
alloys (1500 F in 1946; 1800 F in 1966) (see Figure
5-2).

Compressor pressure ratio is rising, single-shaft
compressors with variable stators having reached 15 and
proposed plants with two-shaft compressors going to 28
by 1976 (see Figure 5-1b).

5.2.3. Materials Available

The extent to which alloys needed for future turbine developments are presently available is indicated in
Figure 5-3. First, second, and third generation refers
to the 1970's, the early 1980's, and the 1990's. Second-
generation blade materials for stationary power plants
will include alloys currently under development for advanced aircraft propulsion systems. Third generation
material properties have been projected by assuming a
20 F per year improvement in material temperature, and
by relying on alloys currently being investigated for
second-generation aircraft propulsion systems. Aluminum-
base coatings have been developed to improve corrosion
resistance, type UC for cobalt-base alloys and Jo-coat

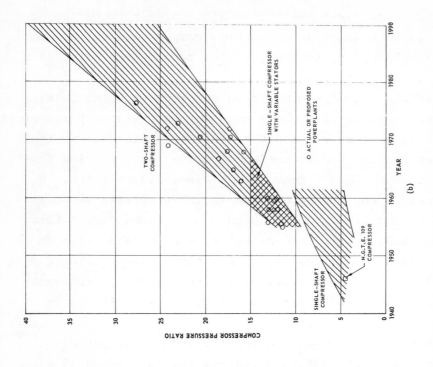

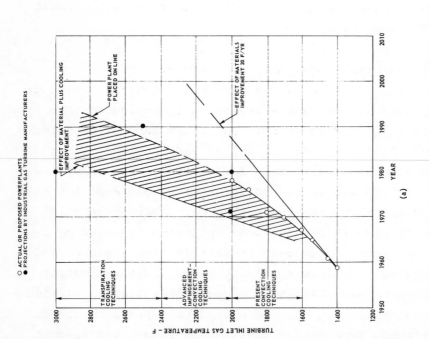

Figure 5-1. Progression of Gas Turbine and Compressor Technology: (a)Turbine Inlet Gas Temperature for Base-Load Operation; (b)Compression Ratio of Aircraft Compressors (Robson, Giramonti, Lewis and Gruber, 1970)

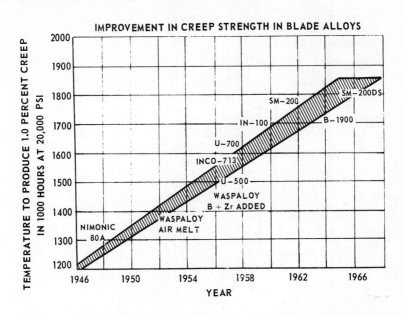

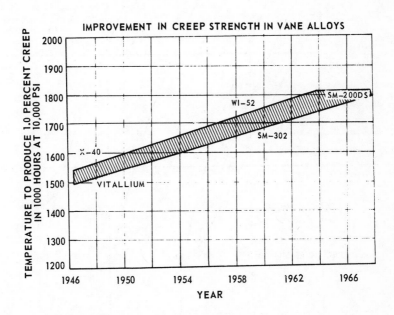

Figure 5-2. Advances in Aircraft Turbine Materials
(Robson, Giramonti, Lewis and Gruber, 1970)

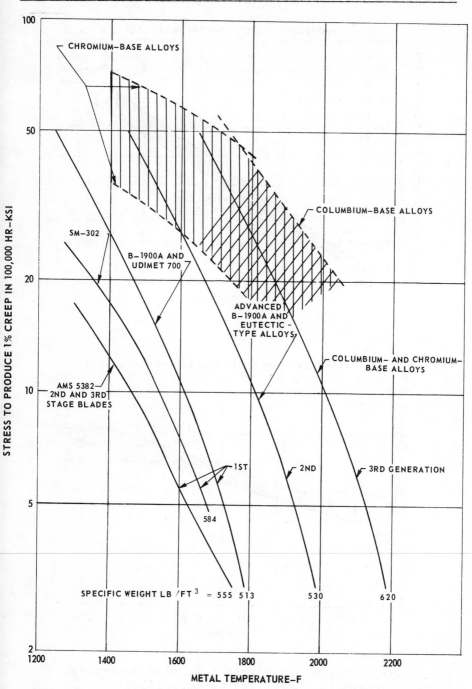

Figure 5-3. Creep Strength for Advanced Turbine Blade
Materials (Robson, Giramonti, Lewis and Gruber, 1970)

for nickel-base; and extension of coating life beyond
the few thousand hours satisfactory for aircraft propulsion is considered feasible (Brancardi and Peters, 1967)
by thickening the coating somewhat. For turbine gas temperatures of 1600 to 1700 F, with air cooling to keep the
metal at 750 F, relatively inexpensive austenitic iron-
base alloys have been used in stationary designs. When
the gas temperature reaches 2000 F the metal will have to
withstand 1200-1400 F; and metals like IN-100 or Incoloy
901 (austenitic iron-nickel alloy) will have to be used.

For compressors, suggested materials are AISI4340 for
low pressure-ratio stages and Incoloy 901 for high, where
rim temperatures might exceed 750 F. For high aspect-
ratio blades AMS 4928 may replace AISI410. Third-generation blades may be fibre-reinforced composites, of which
the two present most promising are silicon carbide-coated
boron fibre and carbon fibre. Borsic®/Ti composites currently being developed should be available for third-
generation industrial units.

For combustors operating above 2000 F turbine inlet
temperature, Hastelloy X or coated TD nickel has been
suggested, and for third-generation systems coated refractory metals could be used. When the combustor is
large the possibility of water cooling always exists.

5.2.4. Turbine Cooling

For inlet temperatures of 1800 F up, blade and vane cooling through several stages will be necessary. Convective
cooling by radial flow through hollow blades or through
round passages disposed around the periphery of a vane
just below its surface will be replaced, as temperatures
rise, by impingement cooling of the leading edge from
inside the blade. Advanced impingement-convection cooling (Figure 5-4, center design) should permit turbine inlet temperatures as high as 2400 F. Film cooling or

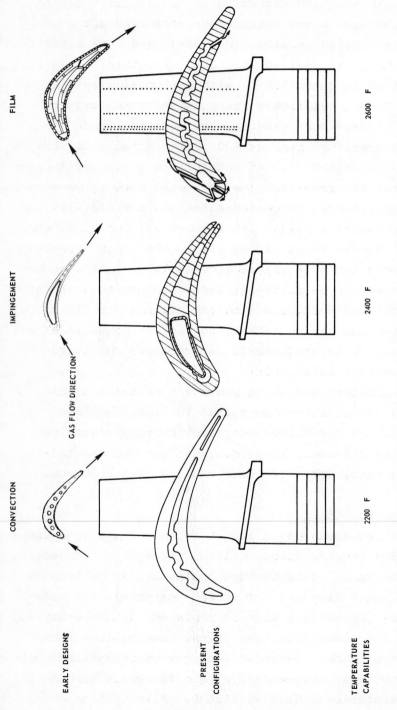

Figure 5-4. Turbine Blade Cooling Improvements (Robson, Giramonti, Lewis and Gruber, 1970)

transpiration cooling [flow from inside through small holes to form a cool boundary layer on the outside (Figure 5-4, right side)], being investigated for aircraft, should be available for third-generation industrial power. The use of sweat cooling--water inside blades and vanes, vaporizing and issuing into the gas stream--has been studied in the past, but not on the blade scale that would be involved in the application under consideration; stress, corrosion, and blocking problems have characterized past efforts. The incentive for the aircraft propulsion industry to study this problem does not exist.

5.2.5. Projections of Base-Load Component Performance
Compressors have climbed in polytropic efficiency from 55 percent in 1940 (Otto-cycle superchargers) to 86% in 1950 to 90% in 1970, with 93% expected (NASA, 1968) (86% is feasible at presently economic pressure ratios per stage.) Current research is directed primarily at increasing the pressure ratio achieved per stage (1.29 in aircraft, 1.12 in industry) without sacrificing efficiency. Tip speeds of 1000, 1100, 1200 ft/sec may be expected in 1st, 2nd, 3rd generation designs, with specific flows of 35-38 pounds/(sec)(ft)^2.

Turbine efficiencies of aircraft units have climbed from 85% in 1950 to over 90% in 1970, with 93% expected in the near future (NASA, 1968; Peters, 1966). It is expected that design change will occur in the direction of higher impulse staging to produce a greater temperature drop in the nozzle and thereby ease the rotor cooling problem somewhat. Turbine tip speeds of present advanced-design aircraft, 1000 ft/sec in the low-pressure last stage to 1600 in the high-pressure stage, are expected to be achievable in industrial designs with the advanced turbine materials referred to above.

5.2.6. Projection of Base-Load Turbine System Performance

The performance of 200 to 500 MW gas turbines operating
on 162 Btu clean sulfur-free gas derived from coal (see
Section 3.2) has been projected, on a time basis referred to as first, second and third generation--the early
1970's, 1980's, 1990's respectively. In each generation
the best material to minimize air cooling-flow requirements was chosen; in no case did the power turbine (the
last stages) require cooling. The computation program
was realistic in its allowance for cooling necessary and
achievable, but understandably simplified in its assumption of a constant flow coefficient per stage and constant mean-diameter flow passages. The cooling requirements were determined from a correlation of fractional
temperature rise — (cooling air to metal)/(cooling air
to turbine gas) — with different cooling techniques,
the correlation having been established from aircraft
experience with advanced impingement-convection cooling
(transpiration cooling required extrapolation of experimental data, but it was not a sufficient improvement to
warrant use). Blade metal temperatures conformed to
constraints imposed by creep strength data (1% in 10^5
hrs, Figure 5-3) applied to blade root stress (Peters,
1970). Table 5-1 below gives some of the constraints
imposed on the calculation of performance of turbines
in each generation. The very busy but meaty Figure 5-5
presents the results, — the effect of compressor pressure
ratio and turbine inlet temperature on the efficiency and
the shaft horsepower per unit air flow rate (SHP/(lb/sec))
for 9 sets of conditions : 3 turbine cooling conditions,
turbine cooling air (a) not precooled, (b) precooled to
250 F, (c) no turbine cooling air used, for each of the
three generations, labelled I, II, III. Consider the
use of unprecooled turbine-cooling air: In the first

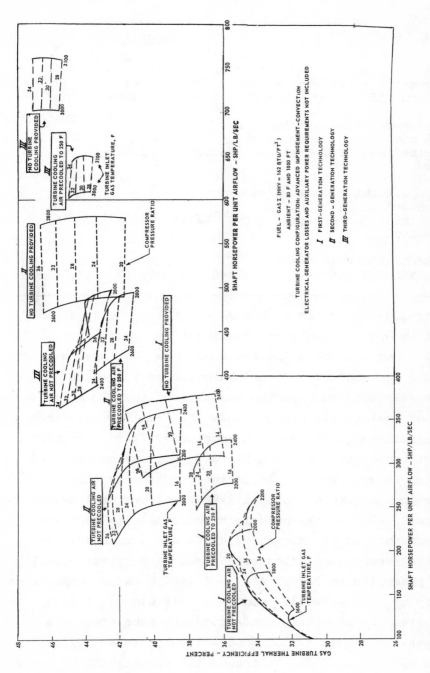

Figure 5-5. Simple-Cycle Gas Turbine Performance with Low Heat-Content Fuels
(Adapted from Robson, Giramonti, Lewis and Gruber, 1970)

Table 5-1.   Conditions of Parametric Study of Base-Load
Turbines

Conditions	Generation		
	I	II	III
Compressor efficiency assumed	89	92	93
Turbine efficiency assumed	90	92	93
Compressor bleed air for turbine cooling is uncooled at turbine inlet temperatures below	2200	2400	2800
Air for turbine cooling is precooled to 250 F if turbine inlet temperature is above	2200	2600	3000
Maximum allowable blade root stress, psi	40,000	40,000	47,000

generation the maximum thermal efficiency of 35.8% is
seen to occur using a turbine inlet temperature of
2000 F and a compression ratio of about 20; and use of
2200 F turbine inlet drops the efficiency because of
losses associated with the greater cooling air needed.
Use of cooling air precooled to 250 F--still the first
generation--yields an efficiency of 37.7% if a turbine
inlet temperature of 2400 F and a compression ratio of
28 are used. Finally, if in the first generation it
were possible to operate at 2200 F without cooling air
(unrealistic because materials not available in first
generation) an efficiency of 40.5% would be possible.
In the second-generation models and with turbine-cooling
air precooled to 250 F, a T_{in} of 2600 F and a compression
ratio of 36 (24) would produce an efficiency of 43.2
(41.2)%. In the third generation the same cooling-air
temperature and a T_{in} of 3000 and compression ratio of
34 would yield 44.8%. Although the calculation of these
system thermal efficiencies is based on realistic
assumptions, the results can be misunderstood. Because

the efficiency of manufacture of 162 Btu producer-gas
from coal is improved by production at high pressure,
the cost of fuel pressurizing has been charged to the
gas-making process; and the gas turbine benefits. When
162 Btu gas is replaced by 1000 Btu gas, the efficiencies
of Figure 5-5 are down 4 to 4.5% (in 40, not in 100) because of the much lower work content of the richer gas
under pressure.

Studies of turbines operating on a regenerative cycle,
with a regenerator pressure drop of 4% equally divided
between the air and gas sides of the heat-exchanger, led
to the conclusion that in gas-turbine power plants employing low-Btu fuels the use of regeneration does not
provide any significant improvement in efficiency or
specific output; and the comparison gets worse the
higher the inlet turbine temperature. If the fuel were
natural gas the improvement could be significant.

5.2.7. Economics of Base-Load Turbine on Clean Low-Btu Gas

From the above parametric study of expected performance
three base-load gas-turbine power plant designs were
chosen hopefully representative of the best practice in
each generation. All three designs had two spools and
a separate power turbine operating at 1800 rpm, but they
differed in capacity, turbine temperature, compression
ratio, etc., as listed in Table 5-2.

Cost analyses of the three base-load gas turbine systems (Table 5-2) and, for comparison, that of a coal-based
steam system were made, all on the basis of an assumed
price of 20¢/10^6 Btu for coal, and of 40.4, 31.4, and
31.8¢/10^6 Btu for 1st, 2nd, and 3rd generation gas-from-
coal (the third generation shows no improvement in cost
over the second because of a recycle compressor needed
for the 3rd generation sulfur-removal system). Table 5-3
gives the plant cost figures that were obtained.

Table 5-2. Design Features of Simple Base-Load Turbine
System, Using 174 Btu Gas

Features	Generation		
	I	II	III
Output, MW	197	331	506
Compressor pressure ratio	20	32	34
Compressor exit temperature, F	922	1075	1084
Temperature into turbine, F	2200	2800	3100
Temperature out of turbine, F	951	1105	1256
Fuel used: Producer Gas of			
HHV, Btu/ft^3	175.8	172.7	172.7
System thermal efficiency*	35.2	41.5	43.6

*Including allowance of 3% for auxiliary, mechanical,
and generator losses.

Power costs were 7.9, 5.4, and 5.4 mills/kWh for the
three generations, vs coal-based steam costs of 6.23,
5.76, and 5.71. Although the base-load gas turbine
appears not to be cheaper than conventional steam until
the second generation of performance is reached, it is
to be remembered that the gas-turbine plant includes
solution of the fuel-sulfur problem, whereas the cost of
eliminating sulfur dioxide from stack gas must be added
to the steam-plant cost. If this is estimated at 0.6 to
1.2 mills/kWh capital cost and 0.5 to 2.2 mills/kWh operating cost (Sec. 2.5), it is seen that even the first-
generation gas-turbine base-load plant is in the running.

Table 5-3. Plant Cost Summary for Simple Base-Load
Gas-Turbine Power Plant (Costs, millions $, based
on 1000 MW plant)

Features	Generation		
	I	II	III
Structures and improvements	5.56	4.29	4.05
Prime movers	32.51	23.58	25.41
Electric generators	9.68	9.60	9.48
Accessory electrical equipment	4.88	4.65	4.70
Miscellaneous	0.22	0.22	
Station equipment	4.03	1.72	
TOTAL	77.5	60.8	61.7

5.2.8 Projection of Performance of Combined Gas Turbine-Steam Turbine Plants Burning Low-Btu Gas

The Russian literature refers to such systems as steam-gas plants, the UAC report (Robson, et al, 1970) as COGAS units. The latter report considers the incorporation of advanced-design features of aircraft gas turbines into COGAS units. Five variations are considered: (1) exhaust-fired, (2) waste-heat recovery (Figure 5-6), (3) conventional supercharged, (4) gas-generator supercharged, (5) two-pressure supercharged. All five were studied in depths differing in detail but sufficient to establish preference. Numbers (4) and (5) were eliminated because they were some 8% below the best. No. (2) is a special case of No. (1), with the secondary fuel firing going to zero. No. (3), conventional supercharging, has been used in Europe with the Velox boiler; and this system would be thermodynamically superior with gas-turbine development stopping at an entering temperature of 1700-1800 F. Furthermore, with this temperature limit it might be possible to burn coal directly in fluidized-bed combustors, though there could be problems

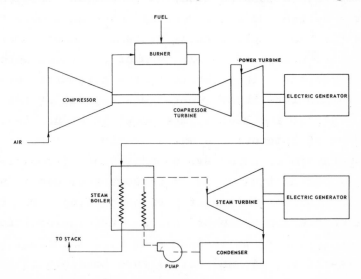

Figure 5-6. Waste-Heat Recovery COGAS System (Robson, Giramonti, Lewis and Gruber, 1970)

of pressure-drop, solids carry-over, and possibly sulfur removal. As entering turbine temperatures rise, however, waste-heat systems will be capable of the same levels of efficiency as supercharged systems, and their capital cost will be lower because of higher fractional partici- pation of the gas turbine in producing power.

Considerations of simplicity and thermodynamic super- iority led finally to the choice of waste-heat recovery as the system deserving detailed parametric study. Realistic allowance was made for pinch temperatures in the waste-heat boiler (50 F, ΔT gas to feedwater at cold end; 100 F, gas to steam at hot end). The boiler heat- transfer surface was based on 8.3" H_2O pressure drop through the boiler on the gas side, though the more compact design associated with a Δp of 50" H_2O would have dropped the system thermal efficiency only 0.7%. Auxiliary power allowances were 0.3% of the gas turbine output +1.7% of the steam output (not the 4.7% charac- teristic of gas-fired steam stations, because forced- draft and induced-draft fans are not necessary. The high-temperature turbine material selection, aerodynamic flow path design and blade-cooling techniques were the same as for the simple turbine systems, though compres- sion ratios, number of spools, number of stages were of course different. The results of the parametric studies, again divided into three generations typical of the early 1970's, 1980's, and 1990's, are shown in Figure 5-7. The effect of compressor pressure ratio and turbine in- let temperature on power system efficiency (electrical output/higher heating value of gaseous fuel used) is shown, for two conditions of turbine cooling. For the case of compressor bleed-off air for turbine cooling precooled to 125 F, it is seen that system thermal ef- ficiencies of 47, 54, and 58% (!) are possible in the first, second, and third generation. [It is to be

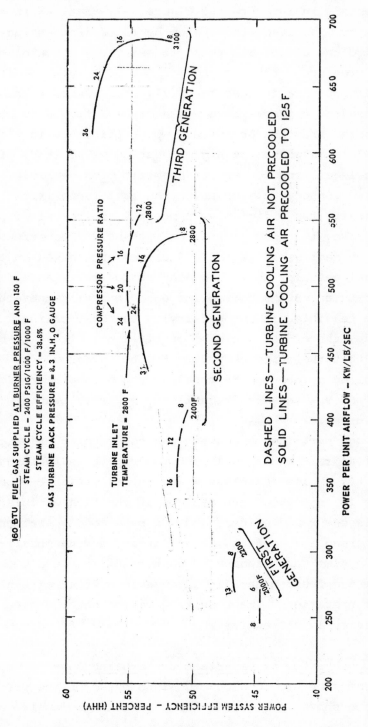

Figure 5-7. Performance of Waste-Heat COGAS Power System (Adapted from Robson, Giramonti, Lewis and Gruber, 1970)

remembered, in interpreting Figure 5-7, that, as in the
simple-turbine case, the system benefits from having the
large volume of low-Btu fuel gas delivered at turbine
inlet pressure with no charge to the power plant. If
natural gas were instead the fuel, the reduced volume
delivered at high pressure would cause 2 to 2.5 to be
subtracted from the ordinate.] The efficiency in Figure 5-7 is based on the higher heating value (HHV) of
the fuel gas, and not on its chemical plus sensible
energy. Consequently, raising the fuel temperature
would raise the efficiency (almost 1% of its value, per
100 F rise, in third-generation systems). The effect
would be negligible if the efficiency were based on total fuel enthalpy above the dead state at a standard
temperature. But since an increase in fuel temperature
may be feasible without penalty to the fuel-making
process, the effect could be significant. It is to be
noted that the efficiency from Figure 5-7 should be
multiplied by about 0.96 to allow for a generator efficiency of 98% and for 2% auxiliary power allowances (see
previous page).

Though part-load operation was not studied, a drop to
half load is roughly estimated to drop the efficiency to
about 80% of its full-load value unless variable-geometry
vaning is supplied. But systems of the high efficiency
of this one would be designed for base-load operation.

Intercooling, reheat, regeneration, and secondary-
steam cycles were studied in connection with the COGAS
system and found generally incompatible with waste-heat
boiler operation. The resulting retention of cycle
simplicity is attractive.

5.2.9. Economics of Gas-Steam Cycle Using Low-Btu Gas

The next step was to choose designs based on the performance shown in Figure 5-7 and to estimate their cost.

Since gas turbine cost per unit output generally falls
with rising turbine temperature, the highest temperatures
allowable were chosen for each generation - 2200, 2800,
3100 F for the three. Optimizing the compression pressure ratio was not so easy; analysis yielded 8, 12, 20.
The steam exhaust pressure was fixed at 2" Hg (101 F),
and no extraction steam was provided for feedwater
heating. Pinch ΔT's in the waste heat boiler of 50 and
100 were chosen for gas-to-liquid and gas-to-steam heat
transfer. Study of a range of steam conditions led to
the choice of 2400 psi and 1000/1000 F as the minimum-
cost system for all turbine-exit gas temperatures exceeding 1100 F— a condition met by all three generations. (The reasonably realistic character of the cost
estimate is indicated by the waste-heat boiler figures;
the study group's estimate was 5% above a later estimate
received from a boiler manufacturer). Table 5-4 gives
significant characteristics of the three generations of
plants.

Table 5-4. Proposed COGAS Power Systems

Generation	I	II	III
Number of gas turbines	3	2	2
Turbine inlet temperature, F	2200	2800	3100
Compressor pressure ratio	8	12	20
% of airflow bled for cooling	4.7	8.5	9.0
Turbine exhaust temperature, F	1297	1514	1485
Compressor-turbine overall length	33 ft	27 ft	26 ft
Single steam turbine, of size	431 MW	381 MW	312 MW
Stack temperature, F	314	219	241
System efficiency*	47.0	54.5	57.7
Total capital cost, Millions	$109.3	$94.0	$89.3

*Electric generator losses and auxiliary power require-
 ments not included. Multiply by 0.96

For estimation of power costs, operating and maintenance costs were assumed 0.2 and 0.5 mill/kWh for steam
power and gas power, respectively; supplies and materials, 0.2 mills; fuel 42, 33, 34¢/10^6 Btu, based on
gas made from coal at 20¢/10^6 Btu. With a 14% capital

charge and 70% operating time, the total cost is 7.3, 5.3, 5.2 mills for the three generations. These are to be compared to 6.3 mills for a first-generation steam plant. [After adding the cost of stack-gas SO_2 treatment for the steam plant, first-generation COGAS is somewhat cheaper than steam. Comparing conventional steam plants with the simple base-load turbine and with COGAS—with the last two based on clean gas from coal—,one concludes that the optimum road to travel is

first generation: COGAS is cheaper than simple gas turbine is cheaper than steam + stack treatment for SO_2.

second generation: COGAS is slightly cheaper than simple base-load gas turbine is significantly cheaper than steam, even with no stack SO_2 cleanup for the steam system.

third generation: COGAS is cheaper than simple turbine is cheaper than steam, even with no stack SO_2 cleanup for the steam system and with COGAS improved to meet the more stringent fuel-sulfur requirement of the 1990's.

Not included in the cost is any charge for thermal pollution. If the thermal efficiency of gas-making is 82, 85, 88 for the three generations, the total efficiency and thermal pollution are as given below.

Table 5-5.   Thermal Pollution of Power-Plant Types

Power System	Steam	Simple Gas T.			COGAS			Nuclear Lt.	
		I	II	III	I	II	III	Wtr	Br
Overall efficiency, coal to electricity									
	38.8	28.9	35.3	38.4	37.0	44.5	48.7	32	40
Net total thermal pollution per unit of electrical energy									
	1.58	2.46	1.83	1.60	1.70	1.25	1.05	2.12	1.50
Liquid thermal pollution, Btu/kWh									
	4530	0	0	0	1910	1705	1400	7230	5110

5.2.10. Other Thermodynamic Cycles

Speculation on and numerical estimates of the advantages
of other heat-engine working fluids than steam and hot
combustion products have intrigued thermodynamicists for
many decades, as has a study of the combination of existing cycles in different ways. Seldom has the cost of
equipment and operation been considered adequately in
such studies. The group responsible for the results reported in Sections 5.2.6-5.2.9 examined a large number
of thermodynamically intriguing combinations, including
a potassium vapor topping cycle, an ammonia bottoming
cycle, and the use of fluorcarbons; and reasons were
given, convincing to the present writers, for concluding
that none of these is competitive with COGAS. Closed-
cycle gas turbine power systems, similarly, were found
inferior to advanced open cycle systems, at least where
use of fossils fuels is involved. The closed-cycle
study included use of He, A, CO_2, SO_2.

5.3. Magnetohydrodynamics

5.3.1. Introduction

In magnetohydrodynamic (MHD) power generation an electrically conducting gas is forced through a duct at high
speed in the presence of a transverse magnetic field.
The electromotive forces induced in the gas allow current to be extracted with appropriately disposed electrodes, and delivered to an external load circuit. The
possibility of using fossil-fuel combustion products as
the conducting gas in large-scale MHD units is of some
interest because such a system, if workable, would offer
thermal efficiencies of 50% or higher. The MHD generator
requires an electron concentration in the gas of about
$10^{14}/cm^3$, which requires very hot combustion gases
(4000-5000 F) seeded with materials having low ionization

potentials (e.g., potassium or cesium). A number of
different types of MHD cycles have been considered, but
the general consensus is that the most likely near-term
hope for using MHD to produce central-station power is
the open-cycle (or once-through) fossil-fueled system
(Bueche, 1969). The MHD unit in such a system would
serve as a "topping" unit, above a steam unit.

A recent Office of Science and Technology panel on
MHD (1969) identified a number of problem areas in which
research and development are needed before work on MHD
power systems can justifiably proceed to a full-scale
prototype. The O.S.T. panel recommended that the U.S.
Government encourage work on the difficult problems of
coal-burning open-cycle MHD systems through annual funding of a research and development effort of about $2
million, and that work on a complete coal-fired MHD
demonstration system should await meaningful success
on these problems. The recommended studies were expected to take 3 years or more. Pursuant to these recommendations, an MHD Task Force, representing both private
and public utilities and governmental agencies, was
appointed to formulate the research and development
program. An M.I.T. study group was contracted by the
Office of Coal Research to help the Task Force formulate
its program. An abstract of the research problems
identified by the M.I.T. group is given in the following
section.

5.3.2. Research Problems (M.I.T. Study Group on MHD,1971)
An assessment of the technology bearing on each component
of the open-cycle coal-burning MHD power generator re-
veals a number of unsolved problems in the areas of coal
combustion for MHD, MHD generators, electrical conductivity of combustion gases from coal, materials problems,
and seed recovery and gas cleaning. These problems are

described below.

Coal Combustion for MHD. A coal combustor is needed
which will satisfy the following constraints of MHD
operation: It must handle U. S. coals having 10% or
more ash; in order to attain suitable efficiency the
temperature of the gases entering the MHD channel must
exceed 4220 F, which requires minimal heat loss from
the combustor; the use of oxygen or oxygen-enriched
air is excluded because of cost; air preheat temperatures must be limited to 2000 F until there is available
either an effective technique of ash removal from the
combustion products or preheaters operative with slag-
laden gas; highly efficient removal of ash or slag from
the gases entering the MHD channel is desirable, but it
must be achieved without an appreciable loss of seed in
the slag; char, which is more desirable than coal for
MHD fuel on account of its lower H/C ratio, is excluded
from consideration until the development of a char-
producing coal gasification industry is definite. The
systems believed to offer potential for eventually meeting these constraints include single-stage combustors,
two-stage combustors, and compact gas-fired combustors
close-coupled with a coal gasifier. The first of these
offers a high probability of fulfilling the design requirements for an MHD unit which can tolerate an ash
carry-over into the MHD duct of 25% or more of the coal
ash. The prospects for further development of the
second and third systems should be evaluated in bench-
scale units.

MHD Generators. The MHD generating channel, which is
the energy extraction device for the high-temperature
portion of the fossil fuel combustion cycle, must
extract 20% or more of the input thermal power (enthalpy)
to the plant at an isentropic component (turbine)
efficiency of 70% or greater if an overall MHD plant

efficiency of 50% or greater is to be attained in a 1500 MW(t) plant. In order to meet these requirements marked improvement in generator performance will be required beyond that which may be achieved simply by scaling up present experimental generators. The most important areas to be studied in any attempts to achieve the required enthalpy extraction and isentropic efficiency of MHD generators are the generator electrical configuration and loading, thermal and viscous losses, electrode and insulator wall breakdown, electrode losses and voltage drops, and the electrical and aerodynamic stability of the generator.

Electrical Conductivity of Coal Combustion Gases. Prediction of the electrical conductivity of the hot gases in MHD power production needs to be achieved with an error less than about 5%, which corresponds to a 5% uncertainty in the power output of the MHD portion of a given plant, or to a 5% uncertainty in MHD channel length. This degree of accuracy is far beyond present capability for dealing with seeded coal combustion products. The main reason for the difficulty is inadequate information on the following factors: the complex composition of the gas, which may be kinetically controlled rather than in chemical equilibrium; slag condensation within the channel, with its possible influence on the potassium concentration; the relatively low channel-exit gas temperatures, at which electron ionization kinetics may play an important role. There is need for experimental measurements of conductivity under conditions that simulate those expected in full-scale generators.

Materials Problems. The severe conditions necessarily encountered in the MHD power generator produce a number of serious materials problems. The conditions include extremely high temperatures (2200-3600 F or

higher, depending upon the component considered),
corrosive environments (the presence of sizable quantities of seed and coal slag in various forms and varying in composition), thermal cycling (over several
degrees in some cases, over several hundred in others),
and long duty cycle (10,000 hours for the MHD duct and
many other key components). In addition, the requirement for the mutual compatibility of different materials
in these environments poses significant constraints on
the materials selection and component design. The
present state of material technology is inadequate in
several respects for the achievement of full-scale coal-
fired MHD power generation. The most difficult problem
areas are those represented by the air preheaters,
electrodes, and insulators, but noteworthy advances in
materials technology are also required for a number of
components, including nozzles, valves, duct work, and
boiler tubes. The following areas should be explored
to provide reliability and wear characteristics and to
find new candidate materials for MHD systems: corrosion
studies (with and without electric fields and currents);
performance studies; small-scale studies on integral
parts; development and characterization of new
materials; and studies of slag characteristics (coal
slag alone and coal slag with seed).

Seed Recovery and Gas Cleaning. Use of even the cheap-
est acceptable seed material)probably K_2SO_4 at about
5¢/lb.) requires seed-recovery efficiencies exceeding
98% if the cost of seed make-up is to be less than
3-10% of the fuel cost. In coal-fired MHD systems, the
seed must be collected and separated from slag before
recycling. Deposition of seed in the steam plant and
air heater before the electrostatic precipitator is a
problem. The emission of sulfur dioxide and nitric
oxide must be controlled. There are indications that

sulfur dioxide control can be combined with seed recovery
by seeding with potassium in sulfur-free form and col-
lecting K_2SO_4. The concentrations of NO in the stack gas
from MHD power plants is expected to be very high (possibly of the order 10,000 ppm) because of the unusually
high temperature of the combustion gases. Stack gas
emission standards of 100 ppm NO may have to be met, but
no gas cleaning systems for NO have yet been developed to
the pilot-plant stage. Small-scale research on gas
cleaning and seed recovery should focus on the determination of NO, NO_2, SO_2, and SO_3 to be expected at MHD
steam plant exit, the measurement of chemical and physical
properties of slag-seed mixture and deposition rates, and
studies of methods for removal of NO_x and SO_x from flue-
gas.

5.3.3. Conclusions and Research Recommendations

The fundamental problems in MHD development--gas conductivity, seed recovery, and materials--remain unsolved
despite more than a decade of effort by numerous
competent research teams in several countries. Now added
to these hurdles are problems introduced by environmental
and fuel constraints, including nitric oxide formation in
high concentrations and various effects of coal slag.
These obstacles leave little room for enthusiasm about
the prospects for MHD development, especially since the
contribution which MHD might eventually make--electricity
from fossil fuel at a thermal efficiency exceeding 50%--
can soon be made by combined gas-turbine steam-turbine
plants (see Section 5.2). The appropriate path to
improving the prospects for MHD central-station power
generation is investment in fundamental, small-scale
research in the critical areas outlined above from the
M.I.T. study group's report. Significant progress must
be made along these lines before research and development

funds are directed to the construction of large MHD hard-
ware.

5.4. Superconducting Electric Generators

In electric power generation the savings consequent on
scale increase in the construction of alternators have
been so great that those devices have become gigantic.
Alternators with capacities up to 1500 megavolt-amperes
and with shafts thirty feet long are presently available,
and the limit on further size increases is shippability.

The magnetic field of a power-plant alternator is
produced in the rotor windings. A decrease in specific
resistance of the winding material would permit an
increase in field strength, a decrease in rotor size and
an increase in power output. To this end a study of the
use of superconducting material and low temperatures in
the field windings of alternators was initiated several
years ago. The results have been so promising as to
merit being briefly presented here (Thullen et al, 1971;
Woodson et al, 1971).

A 45 kVa superconducting-field alternator--a rotating
Thermos bottle full of windings--has been built and
tested, and a 2000 kVa unit only 5 3/4" in external
diameter is under construction, for operation at 3600RPM.

Turbine-generator sets cost in the neighborhood of
$15/kVa, of which the generator may cost one-half or less.
It is estimated that a 1000 MVa superconducting-field
alternator can be built for $6 to $7.5/kVa. The field-
winding loss of such a unit would be only 1 to 2 MW in
1000, with refrigeration consuming only tens of kilowatts.
The use of superconduction (the wire now costs $1/1000
ampere-foot) allows the magnetic field to be upped from
15 to 30 kilogauss or more, possibly by a factor of 3
over conventional alternators; more conductors can be
put in the stator because the iron teeth are eliminated;

lower voltage exists between bars, resulting in less insulation, more copper and 4 to 8 times the power output.
Further, the electrical characteristics of the machine
are different, and it is easier to control.

With easier rotor cooling and much less cooling to do,
the superconducting-field alternator looks very attractive as generator unit capacity continues to climb. The
research is presently financed by the Edison Electric
Institute.

5.5. Fuel Cells for Central-Station Power

Fuel cells convert the chemical energy of fuel directly
to electrical energy without fundamental limitations on
efficiency such as those imposed on heat engines. The
fuel and oxidizing agent are continuously and separately
supplied to the anode and cathode, respectively, at which
electrochemical reactions occur which supply power by
driving current through a loaded circuit. The concept of
supplying large-scale power by fuel cells has received
attention because, in theory, fuel cells have potential
for highly efficient fossil fuel utilization with minimal
effects adverse to the environment. In practice, however,
fuel cells do not give the cost and performance characteristics demanded in commercial application, and they are
now used only in areas where cost is of minor importance,
such as military and space applications (see, for example,
Pouli, 1970). The technological status of fuel cells for
central-station power generation is illustrated in the
following brief discussion of two types of cells that
have been studied for that application.

The molten carbonate fuel cell is probably one of the
most likely candidates for power-plant use. These cells
use an electrolyte consisting of a binary or tertiary
eutectic of lithium, sodium, and potassium carbonates,
and operate at 500-750 C. One of the main incentives

for using this electrolyte is its compatibility with CO
and CO_2 in gases generated from fossil fuels. Catalysts
are generally required in spite of the high temperatures.
Nickel is usually used as the anode material but silvered
zinc oxide is also suitable. Materials used as the
cathode include silver and lithiated nickel oxides. Hydrocarbon gas can be used directly if care is taken to
avoid coking from cracking reactions.

A detailed engineering design study of a large-scale
molten-carbonate fuel-cell power plant, performed in 1962
by the Central Electricity Generating Board in Great
Britain, indicated that a fuel-cell generating station is
not capable of achieving economic parity with steam-
turbine plants. A more recent examination by the Institute of Gas Technology (Ng et al, 1970), of the technical
and economic aspects of a 400 MW molten-carbonate fuel-
cell power plant identified a number of areas in which
further research is needed if a commercially viable fuel-
cell generating station is to be built. Troublesome areas
yet to be solved include cell performance, mechanical
limits of size and configuration of the electrolyte brick
and techniques for its manufacture, and material technology of cell components.

The other example of fuel cells directed to power-plant
use is the solid-electrolyte system studied in Project
Fuel Cell during 1962-1970 (Westinghouse Electric Corporation, 1970). The electrolyte is a thin (20 microns)
film of oxide ceramic (zirconia) stabilized in the cubic
fluorite crystal structure. At the operating temperature
of 1870 F, oxygen ions ($O^=$) flow through the electrolyte
by activated diffusion. The anode is metal (cobalt or
nickel) made porous by the inclusion of a ceramic
skeleton, and the cathode is electronically conductive
oxide such as $(In_2O_3)_{0.98}(SnO_2)_{0.02}$ fabricated in a
porous structure readily permeated by oxygen from air.

Fuel gas flowing over the anode reacts with the $O^=$
diffusing through the electrolyte to form H_2O and CO_2,
with the liberation of electrons and thus useful current.
The rate-limiting process at the anode is countercurrent
diffusion of fuel and reaction products. Oxygen accepts
electrons at the cathode where the rate-limiting step is
diffusion of oxygen through the electrode structure. Gas
from either side does not penetrate the electrolyte
layer, and reaction can only occur when current is drawn.
The objective of the studies which used this fuel cell
was the development of a commercial fuel-cell power-
generating system which would use coal as fuel. The coal
would be gasified at 1750 F, utilizing heat produced in
the fuel cells by resistance and polarization losses and
oxidation of the fuel gases. Cells with 0.1 kW capacity
were operated; a 100 kW plant was designed but never
built, and the project is now terminated. A number of
problems with each component of the cell and with the
fabrication process for making the complete fuel cell
must be solved before a commercial plant can be realized.
These problems include the development of a porous material,for mechanical support of the combined anode-
electrolyte-cathode layers, which has acceptable impurities and impurity levels, uniformity of porosity, and
is not too expensive; the development of a fuel
electrode with uniform characteristics and capability
of operation over required ranges of fuel composition
and temperature; the development of improved techniques
for forming gas-tight, pore-free electrolyte films which
are less than 20 microns in thickness and stable in
continuous and intermittent fuel-cell operation; and
the development of a suitable technique for cathode construction (the presently used electrochemical method for
making the necessary pores in the chemical-vapor-
deposited indium oxide film is unsatisfactory).

There is no indication that the outstanding problems involved in using fuel cells for central-station power production are close to solution or that significant advancement would result from large spending on research and development at this stage. More progress should be achieved through small-scale, fundamental research before large-scale development is pursued.

References

Biancardi, F. R. and Peters, G. T., 1967. Status Report
on the Utilization of Low-Cost Fuels in Gas-Turbines,
United Aircraft Research Laboratories Report, October,1967.

Bueche, A. M., 1969. "An Appraisal of MHD-1969", prepared
for MHD panel of the President's Office of Science and
Technology, Washington, D. C., February,1969.

M.I.T. Study Group on MHD, 1971. "Open-Cycle Coal-Burning
MHD Power Generation - An Assessment and a Plan for Action",
Report to the U.S. Dept. of the Interior, Office of Coal
Research, June, 1971.

NASA, 1968. "Selected Technology for the Electric Power
Industry", Proceedings of NASA Conference at Lewis
Research Center, September, 1968.

Ng, D.Y.C., Maru, H.C., Feldkirchner, H.L., Baker, B.S.,
and Cochran, N.P., 1970. "An Engineering Study of Fuel
Cell Power Supply for Electrothermal Stage of the Hygas
Process", IECEC Conference, Las Vegas, Nevada, September 20-25, 1970.

Panel on MHD, 1969. "MHD for Central Station Power
Generation: A Plan for Action", prepared for the
Office of Sci. and Tech.by Panel on MHD, June, 1969.

Peters, G.T., 1966. Reference Handbook of Prime Mover
Characteristics - Section I. Gas Turbines, United
Aircraft Research Laboratories Report, 1966.

Peters, G. T., 1970. Advanced Design Gas Turbines for
Base-Load Electric Power Generation, United Aircraft
Research Laboratories, May, 1970.

Pouli, D. P., 1970. "Fuel Cells: Today and Tomorrow",
Heating and Air Conditioning, p. 102, September, 1970.

Robson, F. L., Giramonti, A.J., Lewis, G.P., and Gruber,
G., 1970. "Technological and Economic Feasibility of
Advanced Power Cycles and Methods of Producing Nonpol-
luting Fuels for Utility Power Stations", Report prepared
for National Air Pollution Control Administration, U.S.
Dept. of Health, Education, and Welfare, December, 1970.

Thullen, P., Dudley, J.C., Greene, P. L., Smith, J. L.,
Jr., and Woodson, H. H., 1971. IEEE Trans., Power App.
and Systems Vol., PAS90, No. 2, pp. 611-619, March-
April, 1971.

Westinghouse Electric Corporation, 1970. Final Report on Project Fuel Cell, Research and Development Report No. 57, prepared for the Office of Coal Research, Department of the Interior, Washington, D. C.

Woodson, H.H., Smith, J.L., Jr., Thullen, P., and Kirtley, J.L., 1971. IEEE Trans., Power App. and Systems Vol., PAS90, No. 2, pp. 620-627, March-April, 1971.

Chapter 6

UTILIZATION-RELATED ENERGY PROBLEMS

6.1. Automotive Power Plants

6.1.1. Introduction

The gasoline-powered automotive engine is extraordinarily good in all respects except for its tendency to pollute the air with unburned carbon monoxide and hydrocarbons and with lead and oxides of nitrogen. Mechanical improvements are continuously being made, including reduction of such contributions to unburned hydrocarbon escape as gasoline evaporation, crankcase blowby, and spills when filling a car, together with design of the engine for more complete combustion. Such changes cannot meet the requirements of the Clean Air Act, and more drastic changes will be necessary. Whether these will take the form of major redesign of the Otto-cycle engine, the development of a different type of fossil-fuel-burning power plant, an electric automobile, or combinations of these cannot now be forseen. This Section will present the present and early-future standards set by the Clean Air Act, a summary of NAPCA's opinion and evaluation of proposed technical approaches to emission control, APCO's (formerly NAPCA) program, discussion of replacement of the Otto cycle by gas turbines, steam engines, and electric drives, and comments on the Wankel engine and the stratified-charge engine.

The Clean Air Amendments of 1970 require that the standards of emission control which previously were scheduled to be met in 1980 must now be met by 1975. Interestingly, requiring this degree of control by 1975 virtually ensures that the standards will be met (if they are met) by an engine of conventional design rather than by a new approach to automotive power. Standards for the near future are outlined in Table 6-1.

Table 6-1. Automobile Emission Control Standards*
(Heywood, 1971)

Case	CO g/mi	CO % re-duced	Hydro-carbons g/mi	Hydro-carbons % re-duced	NO_x g/mi	NO_x % re-duced	Partic-ulates g/mi	Partic-ulates % re-duced
Uncontrolled vehicles	125	0	16.8	0	6	0	0.3	0
Present controls	47	62	4.6	73	–	0	–	0
1972 standards	39	69	3.4	80	–	0	–	0
1973 standards	–	69	–	80	3.0	50	–	0
1975 standards**	4.7	96	0.46	97	–	50	0.1	67
1976 standards	–	96	–	97	0.40	93	–	67

* The figures on grams/mile (g/mi) are an average based on a typical urban driving cycle.

** These standards correspond to 1% of fuel C as CO, and 1/4% as HC's. More recently EPA issued final 1975 standards which limit CO to 3.4 g/mi and HC to 0.41 g/mi (Chem. Eng. News, 1971).

6.1.2. Possible Solutions to the Pollution Problem of Conventional Internal Combustion Engines

Table 6-2 presents NAPCA's (now APCO) evaluation of different approaches to reducing emissions from conventional internal combustion engines, and gives the time and cost required to develop each approach, the probability of reducing emissions to the required level, the technical feasibility of doing it, and the order of desirability. The identification of goal levels is:

Goal Level	Emission Limit, g/mi CO	Emission Limit, g/mi NO_x	Hydrocarbons
1	17.25	1.35	1.25
2	11.50	0.95	0.60
3	4.70	0.40	0.25

It should be noted that level 3 corresponds to emission levels roughly equal to the 1975 standards.

Table 6-2. Summary of Opinion and Evaluation, by NAPCA, of Proposed Technical Approaches to Control Emissions from Automobiles and Other Four-Stroke-Cycle, Spark-Ignited Engines (NAPCA, 1970)

Technical Approach	Research		Probability of achieving research goal level, % and, in ( ), technical feasibility[a]			Annual Cost per car $	Order of desirability[b]
	Cost, $ millions	Time, Years	1	2	3		
HC and CO control in vehicle exhaust							
Secondary combustion							
1. Exhaust manifold reactor-fuel rich	5-10	2-6	50-100(E-G)	50-100(E-G)	0(F-P)	40-50	3
2. Exhaust manifold reactor with exhaust recirculation	15	3-6	80-90(E-G)	80-90(E-G)	80-90(F-P)	45	3
3. Exhaust system reaction with exhaust recirculation	2-4	3-6	30-75(G)	30-75(G-F)	-(P)	15	3
Engine factors							
4. Lean engine operation	2.5-10	5	50-90(G)	0(P)	0(P)	0	1
5. Engine refinement	5-30	Continuous	50-80(G)	0(P)	0(P)	5	1

Technical Approach	Research Cost, $ millions	Research Time, years	Probability of achieving research goal level, % and, in ( ), technical feasibility[a] 1	2	3	Annual Cost per car $	Order of desirability[b]
Catalytic oxidation							
6. Oxidizing catalytic system, non-leaded fuels	3–10	4–5	20–80(G–F)	20–80(G–F)	0 (P)	40	4
7. Dual catalytic system, oxidizing, reducing	3–5	5–6	20–75(G–F)	20–75(G–F)	0 (P)	45	4
8. Dual catalytic system: reducing agent	5	5	25 (F)	25 (F)	0 (P)	35	5
9. Oxidizing catalytic system, leaded fuels	2	3	0–90(F–P)	0–80(P)	0 (P)	20	3
Ignition and fuel control							
10. Fuel modifications, additives or alternate fuels	1.5–3	3–5	10–90(F)	0 (P)	0 (P)	15	2
11. Ignition system modification	N/A	N/A	0 (P)	0 (P)	0 (P)	–	1

Table 6-2. Continued

Technical Approach	Research Cost, $ millions	Research Time, Years	Probability of achieving research goal level, % and, in ( ), technical feasibility[a] 1	2	3	Annual cost per car $	Order of desirability[b]
NO$_x$ control in vehicle exhaust							
Exhaust gas recirculation							
1. Exhaust recirculation	2-5	2-3	0-90 (E)	0 (P)	0 (P)	7.5	2
2. Exhaust manifold reactor with exhaust recirculation	15	3-6	80-90 (E-G)	0 (P)	0 (P)	45	3
3. Exhaust system reaction with exhaust recirculation	3	6	20 (G-F)	- (P)	- (P)	15	2
Engine air-fuel ratio factors							
4. Lean engine operation	2.5	5	50 (G-F)	0 (P)	0 (P)	0	1
5. Exhaust manifold reactor-fuel rich	2-12	3-5	0-100 (G)	0 (P)	0 (P)	0-40	2

Technical Approach	Research		Probability of achieving research goal level, % and, in ( ), technical feasibility[a]			Annual Cost per car $	Order of desirability[b]
	Cost, $ millions	Time, Years	1	2	3		
Catalytic reduction							
6. Dual catalytic system: reducing, oxidizing (non-leaded fuels)	3-5	6	20-75(F-P)	20-75(F-P)	0 (P)	40	4
7. Dual catalytic system: reducing agents	5	5	25 (F-P)	25 (F-P)	0 (P)	35	5
Fuel modification and engine refinement							
8. Fuel modifications, additives, or alternate fuels	No data developed; fuel additives may be possible approach to NO_x control.						
9. Engine refinement	N/A	N/A	0	0	0	5	1

a E = excellent; G = good; F = fair; P = poor.
b 1 = most desirable; 2 = second most desirable, etc., to 5.

CO and HC Reduction. NAPCA is optimistic about secondary combustion (thermal reactors) as a means of reducing
CO and HC. The high temperature exhaust manifold approach
uses a rich mixture and adds air to promote oxidation of
CO and HC. The principal difficulty is the need to accommodate high temperatures, requiring special materials or
control devices. A second approach is to use a lean mixture and conserve heat to promote oxidation in the exhaust
system. This gives better fuel economy and avoids the
temperature problems of manifold reactors, but has problems related to lean running and conservation of exhaust
heat. This approach is judged less likely to succeed in
reducing emissions than the reactor approach.

Modifications of traditional engine design parameters
with emissions in mind are considered a desirable approach
because of their simplicity and reliability, although they
seem unlikely to reduce emissions below level 1. Table
6-3 indicates the effect of various modifications on emissions.

Table 6-3. Engine Design Characteristics for Low Emissions
(Automobile Engineer   , 1970).

	CO	HC	NO$_x$
Air:fuel ratio	15:1	15:1	<13:1 or >17:1
Ignition timing	--	retard	retard
Manifold depression	--	below 15 in.Hg*	high
Air inlet temperature	--	high	not above 150-300F
Coolant temperature	--	high	low
Compression ratio	preferably low	low	low
Surface:volume ratio	--	low	assumed high
Exhaust temperature	high	high	--
Combustion chamber deposits	avoid	avoid	avoid
Valve overlap	--	minimum	--
Mode of operation in which quantities of pollutants are high-est	not signi-ficant	decelera-tion and idling*	large loads at low speeds

*
These constraints are avoidable by more sophisticated
carburetion.

The feasibility and desirability of using catalytic
oxidation to reduce HC and CO emissions are considered
only fair. Temperature problems are considered similar
to those for reactors, and catalyst durability remains a
problem, though unleaded fuels help. It is thought that
catalytic converters might be useful in conjunction with
other control techniques.

Fuel modification is regarded as having limited potential, though it is of course applicable to old as well as
new cars. Ignition modification is regarded as having so
little potential that work in this area is not to be
supported.

As far as engine modification is concerned, mechanism
studies are most needed. Most engine research is conducted on a cut and try basis. With a clearer understanding of the physical processes involved in the generation
of pollutants, the cuts would be better and the number of
tries less. These studies should involve understanding
the cylinder-piston fluid mechanics, scaling and modeling
problems and chemical reaction kinetics under engine conditions. As an example of a possible contribution from
fluid mechanics, a significant portion of the unburned
gases comes from hydrocarbons in the quenched boundary
layer, which tends to be rolled up along the walls by the
piston on the exhaust stroke. If the vortex containing
these hydrocarbons could be retained in the cylinder in
the clearance volume and thereby made to join the next
cycle, hydrocarbon emissions should thereby be reduced
materially.

NO_x Reduction. Exhaust gas recirculation to alter air:
fuel ratio and reduce peak cycle temperature is considered
to be the most promising means of reducing NO_x. However,
this approach has undesirable interactions with some of
the secondary combustion approaches to the HC-CO problem,
and there are problems with variation of air:fuel ratio
and recirculation control to meet different conditions

of power and speed. This system in combination with
secondary combustion is regarded as the most promising
overall HC-CO-NO$_x$ system.

Modification of fuel mixture is thought to be a
reliable and simple way of reaching the first level of
reduction but has little potential for lower levels. The
two approaches are lean operation for HC-CO-NO$_x$ control
and rich operation to reduce NO$_x$ followed by an exhaust
manifold reactor for HC and CO.

Use of catalytic converters to control NO$_x$ is regarded
as having only fair to poor potential due to catalyst
durability problems and lead problems, though the latter
can be handled by unleaded fuel. The two approaches
under consideration are fuel-rich operation followed by a
reducing catalyst followed in turn by an oxidizing catalyst with secondary air injection, and use of a selective
reactant such as ammonia for reducing NO$_x$ under conditions favoring CO-HC oxidation.

Fuel modification and engine refinement are not considered to have major potential for reducing NO$_x$. Engine
design considerations which contribute to reducing NO$_x$
were outlined in Table 6-3.

Two major engine re-design possibilities have some
promise for reducing the NO$_x$ to better than level 1 of
Table 6-2, a stratified charge (using fuel injection)
and a precombustion chamber. Both these constitute major
changes in the current engines and need both tests and
back-up studies to see where these changes lead and what
potential they have.

Fuel Modification. A recent survey of fuel modification
as an approach to reducing automobile pollution
was made by a panel under the auspices of the Department
of Commerce (1971). The panel concluded that the Environmental Protection Agency should be concerned with the
following aspects of fuel composition: lead, detergents,

and other additives; aromatic and olefin content; and
volatility.

Lead alkyls have been added to gasoline because this
is the least expensive way to increase the octane rating
of gasoline to the desired levels. Also, lead serves as
a high-temperature valve lubricant. On the other hand
lead compounds in exhaust gases poison the catalysts in
the reactors which will be used on automobiles to meet
the 1975 pollution standards. In addition, lead contributes to particulate emissions from automobiles, fouls
spark plugs, and promotes muffler deterioration. Finally,
lead is toxic, and its distribution through the environment should be discouraged.

The Commerce panel found that it would be impossible
to remove the lead from gasoline immediately while maintaining present octane levels (94 for regular, 100 for
premium). However, the capacity now exists to meet limited demand for unleaded or low-leaded 91 octane fuel.
Capacity to produce large quantities will require additional facilities, with a construction lead time of about
24 months. It was estimated that the retail cost of
unleaded 91 octane fuel would be about one cent more per
gallon than present leaded regular. This aspect of fuel
cost would probably be overshadowed by decreased fuel
economy resulting from the operation of emission control
systems. The panel recommended that EPA establish regulations to ensure the availability of low-leaded and unleaded fuels in a timely manner and in sufficient quantities, and in addition encourage automobile manufacturers to institute a moratorium on increases in octane
requirements in order to eliminate the need for refinery
investment which might otherwise go into increasing the
output of unleaded fuel.

Another area of concern related to fuel composition is
the aromatic content of fuels. Attempts to produce high-
octane unleaded fuels are likely to require significant

increases in the aromatic content of fuels, and this
could increase the photochemical reactivity of exhaust
gases. The panel found the evidence of such increased
reactivity questionable. If high-octane unleaded fuels
are to be widely used, research should be undertaken to
determine whether increasing the aromatic compounds in
gasoline will increase exhaust photochemical reactivity.

Polynuclear aromatic hydrocarbons (PNA's) are thought
to be carcinogenic and they are consequently of concern
as exhaust emissions. PNA's increase as aromatics increase in leaded gasoline, but appear to fall when lead
deposits are eliminated. Research is now under way to
determine what can or should be done about PNA's. Since
the automobile accounts for only 2 to 10% of national PNA
emissions and catalytic exhaust treatment systems reduce
PNA's, the panel did not make a recommendation on this
matter.

The amount of evaporative emissions and the degree of
photochemical reactivity which they exhibit are affected
by fuel volatility and the amount of olefins present,
respectively. However, increasingly effective devices to
reduce evaporative emissions are being introduced on automobiles, and consequently the panel felt that efforts to
control volatility and olefin content would be of little
value except in specific areas experiencing severe smog,
and then only as an interim measure.

Concerning detergents and gasoline additives, the panel
supported their use insofar as they could be demonstrated
to reduce pollution. No attempt was made to discuss specific additives. EPA is now in the process of registering
fuel additives.

6.1.3. Alternatives to the Conventional Internal Combustion Engine

Figure 6-1 shows comparative specific energy (roughly corresponding to range) and specific power (roughly corres-

ponding to speed) for several different power plants. A
good engine is high and on the extreme right of this figure and not too expensive.

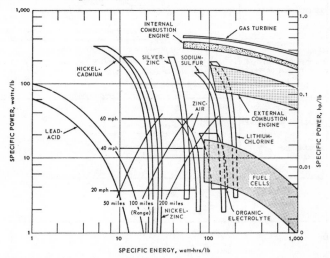

SPECIFIC ENERGY, watt-hrs/lb

Figure 6-1. Vehicle Requirements and Motive Power Source
Requirements (Assumes 2,000 lb vehicle, 500 lb motive
power source and steady driving; power and energy taken at
output of conversion device)(NAPCA, 1970)

A variety of engines are being explored as alternatives to the internal combustion engine. The Air Pollution Control Office of the Environmental Protection Agency
has chosen to concentrate on five systems in its Advanced
Automotive Power Systems Programs: gas turbines, heat
engine/electric hybrids, Rankine cycle engines, heat
engine/flywheel hybrids, and all-electric systems (Figure
6-2). Other systems which have received attention include
rotary combustion (Wankel) engines, fuel cells, stratified-charge engines, Stirling engines, and free-piston
engines. Much work in this area is also being sponsored
by private industry.

Gas Turbines. The basic gas-turbine or Brayton cycle
consists of induction and compression of air by a compressor turbine, introduction of the air into a combustion
chamber at high pressure, ignition and combustion of fuel,
creating a high-pressure high-temperature gas stream which

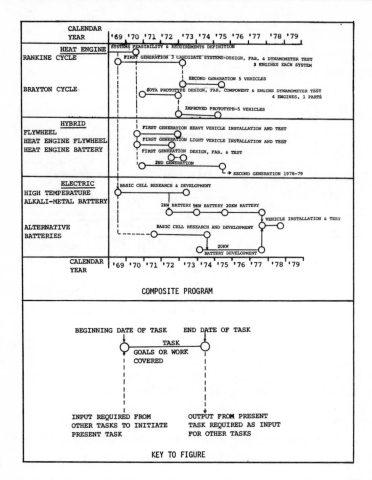

Figure 6-2. APCO Advanced Power Systems Program (Broga, 1970)

expands through a turbine to drive the compressor and provide power through an output shaft. (For a diagram see Figure 5-6.) More complicated turbines employ separate turbines for driving the compressor and for power output, regenerators to recover exhaust heat and improve thermal efficiencies, and other refinements.

Table 6-4 presents emission data for the Chrysler Turbine Car, an experimental model which Chrysler lent to a number of American families for testing starting in 1963. Although the gas turbine offers a clear advantage over the ICE on emissions, it is neither economically nor

operationally competitive in its present form. The oper-
tional problems of the turbine stem from the difficulty
of matching a high-temperature, high-speed powerplant to
a variable load. Durability and fuel economy are the
principal problems which remain to be solved in the application of turbines to automotive use. The economic problems associated with turbine production are also significant. Due to the high speeds and temperatures characteristic of turbines, they require greater precision in

Table 6-4. Emission Data for Chrysler Turbine Car— Cold-
Start, Composite-Cycle, Dynamometer Tests (NAPCA, 1970)

Basis	Exhaust Emissions		
	HC	CO	NO_x
Pounds per mile (NDIR[a])	0.0020[b]	0.0155	0.0041
Pounds per mile (HC by FID[a])	0.0036[b]		
Grams per mile - gas turbine	0.91	7.03	1.86
Grams per mile - gasoline engine equipped with 1968 exhaust emission control system (NDIR)	3.43	35.10	6.76

[a] Abbreviations indicating type of sampling equipment:
NDIR - non-dispersive infrared analyzer, FID - flame
ionization detector.

[b] Hydrocarbon emissions measured by FID and converted to
MDIR using factor of 1.80.

assembly and more high-cost materials than do conventional
engines. In addition, the large scale introduction of turbines would make obsolete most manufacturing and maintenance facilities now used for the automobile, and petroleum refining facilities would probably have to be modified. APCO expects the industry to demonstrate the turbine as an automotive powerplant, which seems likely since
the industry has more experience with the turbine than any
other alternative to the ICE. Consequently APCO is spon-
soring research aimed at solving the problems associated

with turbine introduction rather than a demonstration
itself, with emphasis on the manufacturing problems. At
present, small gas turbines for aircraft sell for around
$25/hp. Introduction of turbine power in buses and
trucks, whose requirements make turbines more appropriate
than is the case for autos, is imminent.

Rankine Cycle (Steam) Engines. Rankine cycle engines operate on the principal of using expanding vaporized working
fluid to operate a piston or turbine. Vaporization of the
working fluid is achieved in an external combustion
chamber. After work extraction from the fluid in an ex-
pander the vapor passes through a feed-water preheater,
then to an air-cooled condenser, then to low-pressure
storage and on through a pump back to the vapor generator. A thermodynamic efficiency of about 24% is achieved.
Figure 6-3 illustrates the cycle.

Most recent developments of this type of engine have
used cylinders as expanders and water as the working
fluid, though there has been interest in turbines and use
of alternative fluids such as freon.

Table 6-5 shows emission data for various recent
Rankine-cycle engines. The driving cycles under which
these data were obtained are not reported, but the performance is obviously very good by present-day standards.

Though several organizations are working on the development of Rankine-cycle engines for automotive use, a
large number of problems remain to be solved. For water
as a working fluid, freezing and quick startup are problems. An important problem is lubricant carry-over,

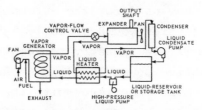

Figure 6-3. Schematic of Typical Rankine-Cycle Steam
Engine Components (NAPCA, 1970)

Table 6-5. Emission Data[a] For External Combustors
Associated with Stirling Engines and Steam Engines
(NAPCA, 1970)

Fuel	Type of Equipment	CO, ppm	HC, ppm	NO_x, ppm
No. 2 diesel	GM Stirling engine- 10 hp[b] with combustion air pre-heater	80	2	500
Kerosene	Williams steam engine	500	20	70
JP-4	Thermo-Electron steam engine	10	--	110
	Steam engine; developer's name[c] withheld by request	3000	30-40	25-35
Diesel fuel	Philips Stirling engine - 80 hp[d] with exhaust-gas recirculation	170	--	38

[a] When the data in this table are compared with those for
gasoline engines on a relative mass-emission-rate basis,
the present data should be roughly doubled to account for
the higher air/fuel ratio generally used with external
combustors. Approximately,1000 ppm of CO, HC, or NO_x is
equivalent to 5 g/mi.

[b] Operated at 25:1 air/fuel ratio.

[c] Data supplied by organization developing engine.

[d] Operated with 50 percent of combustion air from recirculated exhaust gases.

causing collection in the condenser or deposition in the
boiler or both. Cost is a continuing problem and cheaper
boiler alloys are needed than the stainless steel that has
been proposed up to now.

Organic working fluid Rankine-cycle engines reduce the
problems associated with the lubricant but have several
of their own. The working fluid is not as stable thermally as water. The size of the system tends to be larger
and temperature differences smaller, and this necessitates more critical condenser design and, to avoid hot
spots, more careful boiler and burner design. Figure 6-1
shows that Rankine-cycle engines do have promising operating characteristics.

APCO's research effort in Rankine engines consists of parallel programs aimed at developing both overall system concepts for the powerplant and better system components. The three systems contemplated are an organic-fluid reciprocating design, a water-based-fluid design, and an organic-fluid turbine design. Some companies will design systems and produce specifications for system components while others will develop components to meet those specifications.

Electric Drive. An electric drive system requires electric motors, a means of controllable power flow, and a source of electricity. The source can be batteries, fuel cells, or a hybrid system in which a fuel cell or engine continually recharges a battery.

The two types of electric drives receiving serious consideration in APCO's research program are the heat engine/electric hybrid and the all-electric battery-powered vehicle. APCO considers the cost of fuel cells to be prohibitive for automotive use.

Under cruising conditions, the hybrid vehicle draws power from the engine, which runs at essentially constant speed. When additional power is needed for acceleration, it is drawn from the battery. In urban areas, the vehicle could operate on the battery alone to reduce emissions. Because the IC or other engine in a hybrid operates at essentially constant speed, reduction of its emissions would be easier than is presently the case.

The most important problem with hybrid vehicles is the complexity inherent in having two separate power systems. However, better batteries are another major need.

APCO's research program for a fully electric vehicle is expected to produce a prototype much later than the other efforts, in 1978. Fully electric (non-hybrid) automobiles of course have no effective emissions, although lead-acid batteries emit hydrogen and oxygen. However, massive use of electric cars would put an increased load

on electric utilities and cause increased pollution from
the power industry, although because this pollution is
centralized it could be more easily controlled. Whether
the hard-pressed utilities could physically increase
capacity to the extent which would be required by widespread adoption of electric cars is another matter.

Table 6-6 compares emissions from internal combustion
engines meeting the 1975 standards with powerplant emissions resulting from powering electric vehicles. The
emissions per mile for electric vehicles were determined
by calculating the coal requirement to generate the car
energy required, assuming efficiency factors as indicated
for each conversion step. Emissions from coal burning
were determined by using APCO emissions factors for uncontrolled plants. (Natural gas-fired electric plants
would have lower emissions, as would coal plants with
control devices.)

Electric power for automotive propulsion can of course
be supplied by nuclear power. Electrification of vehicular transportation is the logical way to shift from
fossil to nuclear fuel as the prime source of energy for
transportation, should this prove necessary or desirable.

Aside from the question of whether use of electric
automobiles would produce a net reduction in pollution,
the most important technical problem is the low specific
energy of present batteries. Tables 6-7, 6-8, and 6-9
outline the energy requirements for various vehicles
and the characteristics of various battery systems which
could be used in vehicles. Silver-zinc, silver-cadmium,
and nickel-cadmium batteries are not included because
their reactants are too expensive to be considered for
general use.

The major objective of APCO's research program for
batteries is the "proof principle" for lithium-sulfur and
sodium-sulfur batteries. This entails demonstrating that
the energy and power densities required can be achieved

Table 6-6. Comparison of Emissions from Powerplants Serving Electric Cars with Those from Internal Combustion Engines (Agarwal, 1971)

| | Electrical System Efficiencies | | | Automobile Emissions, gram/mile | | |
| | Efficiency Range % | | | Directly From Internal Combustion Engine (1975 H.E.W. goals) | From Generating Station Supplying Electric Car | |
Process			Pollutant		Small: 0.2 kWhr/mile =0.38 lb coal	Full Size: 0.5 kWhr/mile =0.95 lb coal
Generating plants	30	40				
Transformers & Conductors	90	95	Hydrocarbons	0.25	0.016	0.04
Battery: charge	85	95	Carbon monoxide	4.7	–	–
discharge	75	85				
Controller	90	95	Oxides of nitrogen	0.4	1.72	4.3
Motor	65	65	Sulfur dioxide*	–	6.5	16.25
Total (product)	10.3	23	Particulates	0.03	0.85	2.1

*Proportional to sulfur content of coal: 2% assumed.

Table 6-7. Energy Density Requirements for Cruise
(Heitbrink and Tricklebank, 1970)

| Type of Vehicle | Range, miles | Constant Speed Cruise | | | Acceleration |
		Velocity mph	Energy Density Whr/lb	Power Density W/lb	Power Density W/lb
Urban car	50	40	25	20	65
Commuter Car	100	60	55	33	70-103
Family Car	200	70	122	43	73-110
Metro Truck	100	40	33	13	40
Urban Coach	125	30	42	11	35

on an elemental cell basis. If this is accomplished, a
program will be initiated to develop 2, 5, and finally 20
kW cells by 1978. As a contingency, metal-air batteries
are also being developed for use in hybrids. This will be
abandoned if an alkali-metal battery is successful at the
2 kW stage.

   In the development of alkali-metal batteries, the development of materials and packaging which will withstand
the high temperatures at which these batteries operate
will be important. In addition, the understanding of the
electrode processes is basic. Mechanistic studies of
these processes leading to physical models is very important. There is a gulf between the chemical kinetics as
understood on an atomic or molecular level and the rate
processes which limit the overall performance of a battery or fuel cell. A major effort is needed in this area.

   In the very long term, electric automobiles have real
promise. There is no known physical law which restricts
their operation to the lowly position in Figure 6-1

Table 6-8. Conventional Ambient-Temperature Battery Characteristics (Heitbrink and Tricklebank, 1970)

Battery	Cell Reaction (discharge→)	Open Cell Voltage	Theoretical Energy Density, Whr/lb	Practical* Energy Density, Whr/lb	Practical* Power Density, W/lb
Lead-Acid	$Pb+PbO_2+2H_2SO_4 \rightleftarrows 2PbSO_4 + 2H_2O$	2.1	80	2-15	80-15
Nickel-Zinc	$2NiOOH+Zn+2H_2O \rightleftarrows 2Ni(OH)_2+Zn(OH)_2$	1.7	170	15-25	100-30
Zinc-Air	$2Zn+O_2+2H_2O \rightleftarrows 2Zn(OH)_2$	1.6	480	40-75	35-20

* Representative Values.

Table 6-9. Advanced High-Temperature Battery Characteristics (Heitbrink and Tricklebank, 1970)

Battery	Cell Reaction (discharge→)	Temp. C	Open Cell Voltage V	Theoretical Energy Density, Whr/lb	Projected*+ Energy Density, Whr/lb	Projected*+ Power Density, W/lb
Lithium-Chlorine	$2Li+Cl_2 \rightleftarrows 2LiCl$	650	3.5	1000	150-200	90-180
Lithium-Sulfur	$2Li+S \rightleftarrows Li_2S$	350	2.3	1220	110-160	250-360
Sodium-Sulfur	$2Na+3S \rightleftarrows Na_2S_3$	300	2.1	360	80-150	90-160

* These values are taken from the literature and it is not known whether they are estimates based on equivalent battery packages.

+ The ranges given for the projected energy and power densities are not to be regarded as being interrelated as is the case in Table 6-8, but they are an indication of the uncertainties in the projected engineered battery systems.

which they now occupy. Much long-term basic research is
needed in this area, however, to develop the promise inherent in electrical systems.

Heat Engine/Flywheel Hybrids. Another area of emphasis
in APCO's program is the development of heat engine/fly-
wheel hybrids. The system is analogous to that of the
heat engine/battery hybrid except that the energy storage
device is mechanical rather than electrical.

Research is based upon the attainment of an energy
density of 30 Whr/lb. Two types of flywheels are under-
going testing to verify predicted energy and power densi-
ties--one made of fiberglass and the other of steel.
Work is expected to begin shortly on a transmission configuration for the system. This system does not have
high priority, ranking after the gas turbine, Rankine-
cycle engine, and heat engine/electric hybrids.

Other Engine Types. There are other possible anti-
pollution automotive engines not currently emphasized in
the APCO AAPS program. One of these is the Wankel rotary
combustion engine illustrated in Figure 6-4. Automobiles
are currently in production with this engine in Japan and
Germany. Due to the characteristics of the combustion

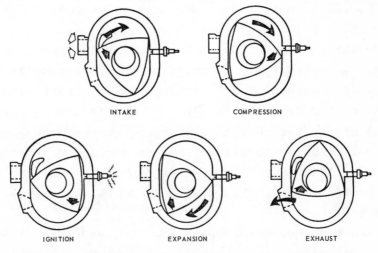

Figure 6-4. Sequence of Wankel Rotary Engine Cycle
Events (NAPCA, 1970)

chamber, the Wankel engine in its basic form has higher
emissions than the conventional piston engine. However,
its exhaust gases are hotter, rendering them more suitable for post-combustion treatment, and the small size
and weight of the Wankel makes it attractive for use in
conjunction with bulky emission control equipment.

Rotor sealing, seal life, and spark-plug life have been
problems in the development of Wankels, but progress has
been made. Manufacturing and servicing of Wankels present
some problems not encountered with conventional engines,
but the dislocations which would be caused by a switch to
Wankels would not be as serious as might be the case with
some other engines.

The stratified-charge engine is similar to the conventional internal combustion engine except in combustion
chamber design, use of fuel injection, and the combustion
process. Fuel is injected into the cylinder as the spark
appears rather than in mixture form regulated by a carbur-
etor. The mixture need be ignitable only near the spark,
and becomes lean further away in the cylinder. The combustion products are similar to those resulting from very
lean engine operation and are very low in HC and CO. More
work is required to reduce NO_x levels, however.

Problems which have been encountered in the development
of the stratified-charge engine include fuel economy and
ignition, particularly in relation to spark plug life.
Fuel coking is believed to cause both the spark plug difficulties and injector fouling. The stratified-charge
engine offers the advantage of using many of the service
and manufacturing facilities now in use for conventional
engines. APCO says that if this engine is added to the
AAPS program, the emphasis will be on NO_x development and
fleet testing.

Another power system which has received some attention
as an anti-pollution alternative is the Diesel engine.
Although it is heavier, costlier, and less powerful than

comparable conventional engines, it has low CO and HC
emissions, and NO_x emissions which are believed to be
within striking distance of the 1975 standards. Work on
the Diesel as an alternative to conventional power systems would concentrate on reducing NO_x and on improved
performance in automobiles.

Other alternative power sources which are receiving
less attention include the Stirling engine, which is
bulky, expensive, and not particularly flexible but might
be suitable for use in hybrid vehicles. APCO will not
support Stirling research itself, but will closely monitor
the progress of the N.V. Phillips Co. in Holland, which is
developing large and small Stirling engines for vehicular
size. Another possible alternative engine is the free-
piston engine, which has emission characteristics similar
to those of the Diesel but costs significantly more, thus
reducing the possibility that it will be the economical
solution to the vehicle propulsion problem.

One of the great difficulties involved in developing
a low-emission vehicle with an unconventional powerplant
is that when it is first introduced it is likely to cost
significantly more than conventional automobiles.* Consequently, it will be very difficult for any new vehicle
to get the foothold in the market which will be required
if costs are to be reduced through refinement and
increased scale of production. To alleviate this difficulty the Clean Car Incentive Program was established to
provide a market for low-emission vehicles by buying
vehicles meeting certain standards at premium prices for
government use.

---

* As an example, the price of new Rankine-cycle automobiles produced in large numbers is difficult to estimate,
but the Williams Co. in 1967 was taking orders for a
steam-powered Chevelle at $10,250 each, in lots of ten.

6.1.4. Research Needs

Occasional reference has already been made to research
needs. These will now be summarized and, to some extent,
amplified.

Basic research needs include the following:

Chemical. Kinetics of combustion in engines, directed
toward a better understanding of what limits the combustion process, how NO_x is formed, how combustion interacts
with radiation and convection in engines, what limits
completion of exhaust-gas combustion.

Kinetics of catalytic combustion, including a continuing search for catalysts not poisoned by lead.

Kinetics of NO_x elimination by dissociation or by
reduction.

Fluid Mechanics. Flow of gases in engine chambers; interaction of flow and combustion; control of burning and mixing patterns.

Heat Transfer. More sophisticated studies of heat transfer and associated power requirement, with particular
application to optimization of air-cooled condensers for
Rankine-cycle engines.

Materials. Fibre research to maximize safe storage of
energy in flywheels.

Electrochemical. Continued search for storage battery
components of high specific power and specific energy,
including analysis of diffusional limitations.

Fundamental studies of which the above are typical
must be pursued to keep options open, but measurable
movement toward a practical solution necessitates
development effort by industrial laboratories, federally
sponsored when the prospects of gain constitute an inadequate incentive. This is the present situation and, as
already indicated, APCO has responded with a concrete
program. Basic research of the type appropriate for
sponsorship by NSF can constitute an effective support of
the developmental effort.

6.2. Space Heating and Cooling

6.2.1. Introduction

About 22% of the nation's energy consumption is for space
heating, substantially all of which is now supplied by gas
and oil; electric space heating is growing but is still
very small. Twenty-two percent of 70 quadrillion Btu's is
a large enough item to deserve attention. As is so often
the case, however, a significant improvement in energy
use for space heating is achievable only as the sum of
many separately small improvements--feasible only if a
national conscience is developed concerning the obliga-
tion to use energy effectively.

Most home heating is accomplished by small furnaces
which, when in adjustment, have the efficiencies listed
below (ASHRAE, 1969):

	Efficiency
Anthracite, hand-fired	60-75%
Bituminous coal, hand-fired	50-60
Bituminous coal, stoker-fired	60-75
Oil and gas fired	70-80

These values, however, are somewhat misleading. Furnaces
tend on the average not to be in good adjustment, thereby
reducing the above figures by 5 to 10 when a furnace operates continuously. In addition, space temperature control
by thermostat is conventionally based on on-off burner
operation. During start-up and shut-down oil burner atomization is momentarily poor and air flow rate quite variable because of stack-draft changes, and soot production
can be high; during shut-down, air continues to be drawn
through the furnace, cooling the firebrick setting as well
as transferring heat away from the furnace-heat-transfer
surfaces. Such intermittency can drop the overall efficiency to as low as 50-60%, or in water heaters even to
30%. If a national conscience is to be developed in the
area of effective use of energy, increased home-heating

furnace efficiency is one area worthy of attention.
Better definition of the problem would come from a survey
to determine, for oil- and for gas-burning furnaces, the
f-E relation, where f is the fraction of the nation's
furnaces operating with an efficiency exceeding E. A
guarantee of 80% average efficiency could be met by new
equipment. High-low rather than off-on burners merit
development, but the impetus to develop them is largely
missing as long as the householder is not energy-
conscious.

6.2.2. Organization of Heating and Air-Conditioning Effort

Research needs and prospects for change resulting from
technically successful developments can best be discussed after presentation of the picture of how the
heating and air conditioning industry is organized.
There is very little vertical integration in this industry. In industrial or commercial building construction
customers ask architects to design buildings; architects
use engineers to design heating or cooling systems; contractors build these systems out of standard components.
In domestic dwelling construction the builder may be the
architect as well; the engineer is generally missing;
the installation contractor's influence on the system is
large, and there may be more than one contractor supplying system components. Economic optimization is substantially never considered in a way specific to the subject
dwelling, though studies in that area may have had indirect effect on decisions. The prospective householder
has a negligible idea of what the true cost of heating or
cooling his dwelling will be; he often thinks he has,
but the information comes often from a not disinterested
party, the electric or gas company or seller of a particular furnace, and is far from complete.

The heating and air-conditioning industry consists of
many small companies and a few large ones; most in the
former category and many in the latter are unwilling or
unable to support research or development on components
or systems except through membership in ASHRAE (The American Society of Heating and Refrigeration Engineers). A
number of committees of that Society are organized to
solicit and review proposals to do work of interest to
manufacturers of heating and air conditioning equipment;
and other committees representing industry, college and
society members disperse research funds obtained by contributions from manufacturers. The results of the
research are well disseminated, and influence both equipment design and industry codes and standards. The U.S.
Bureau of Standards Building Research Division is also
helpful in supplying new data.

Two general types of problems need study, those concerned with obtaining design or performance data for
devices or materials of present interest to the industry
and those concerned with systems too wide in scope to be
the specific concern of any sector of the industry as it
now exists in unintegrated form. Research needs and opportunities in these two categories will be discussed
separately.

6.2.3. Material and Equipment Development

The building industry is in process of slow but continuous
change in materials and methods of construction, and there
is in consequence a continuing need to determine the
thermal characteristics of new insulator materials, wall
assemblies, etc. New industrial fabrication techniques or
processing operations are developing, and there is need to
determine the associated cooling load which an air conditioning system must handle. Duct shapes and fabrication
and assembly techniques change and new arrangements of
heat-transfer coils or finned tubes are invented; and

pressure drop data or heat-transfer coeeficients must be
obtained. No one of these research or development or
measurement needs is critical, but in toto they permit the
heating and air-conditioning industry to maintain a slow
growth in effectiveness.

6.2.4. System Studies

A good case could be made for the assertion that more improvement in heating and air conditioning--meaning more
and better controlled heat or cold per dollar, without
undesirable byproducts--will come from better use of what
we now know about components and processes than from discovery of new ones; and the improvement will come as a
result of system studies. Systems studies can in part compensate for the absence of vertical integration in the
subject industry or for the financial inability of the
small manufacturer to support a study to decide how best
to recommend use of his product; they can supply a market
analysis that discloses a heretofore unsuspected business
opportunity; they can affect public policy in the energy
area. It is in the area of systems studies, and the wide
distribution of those findings to both the purveyor of
equipment and the consumer, that research in heating and
air conditioning is in greatest need of being strengthened. Domestic heating will be taken as an example of
need and opportunity for systems studies. To simplify the
problem of semi-quantitative discussion of interrelations
existing in this area, the general quantitative statement
of heating cost will first be presented.

Home Heating-Cost Formulation. To save time in exposition
the quantitative relation will be presented first; hopefully, with terms defined, its structure will become obvious.

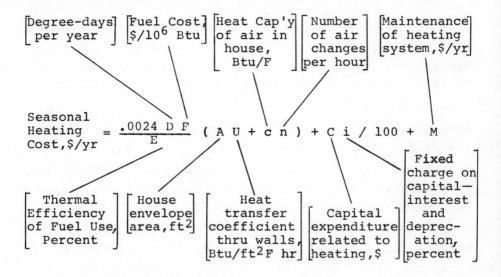

The seasonal heating cost is the sum of three terms,
fuel cost, capital cost, and maintenance. Fuel cost involves D, the degree-days that characterize the locale and
the heating season—the addition, throughout the season,
of the daily amounts by which the outdoor temperature is
less than 65F; F, the fuel cost in $/10^6 Btu, net heating
value (oil at 18.5¢ gal, natural gas at $1.30/1000 cu ft
and electricity at 0.48¢/kWh all correspond to $1.40/10^6);
E, the thermal efficiency of fuel use, in percent; A, the
house envelope area, sq ft; U, the average heat transfer
coefficient of heat loss through the house walls and win-
dows, Btu/ft^{2}hr F (actually, $AU \equiv \sum_j A_j U_j$, where the A_j's
represent the various different kinds of area composing
the envelope); c, the heat capacity of the air contained
in the house, Btu/F (= 0.018 x house volume, cu ft); and
n is the number of air changes per hour, depending on
family living pattern, weather stripping, storm windows,
etc. The next item is C, the capital expenditure on the
complete heating system, house insulation, storm windows
and doors, weather stripping, or any item added for its
heat-saving characteristics; and i, the interest plus
depreciation on the capital investment. The final item is

M, the annual cost of maintenance of the heating system
plus any difference in insurance rates between two possible systems.

It is clear that for a house of specified shape and
size (A,c fixed) in a specified location (D fixed), the
controllable items in a finished house are E and M;
during construction, fuel type F, wall insulation and
air change frequency U and n, and--affected by these--the
capital charge C, are all controllable. A simple example
of use is optimization of expenditure on insulation. The
U obtainable from various expenditures C_I/A ($/sq ft),
where C_I is the capital expenditure on insulation, is
first obtained. The proper expenditure is that one for
which $dU/d(C_I/A) = -i\ E/0.24\ D\ F$.

The above comments should not be interpreted as sug-
gesting that the problems of domestic heating are solved
or solvable by simple application of the equation given.
Installed insulation tends to be poorer than handbook
values; furnace efficiencies tend to be lower than expected, and depend in a complex way on burner capacity,
temperature interval of the thermostat on-off setting,
interaction of stack draft and air-fuel ratio, time since
last furnace cleanout; the relative contributions of
window area, unheated attic floor area, and house wall to
the term AU vary greatly. Generalizations ready for
customer guidance come with difficulty and often only when
modified by field surveys.

Examples of Systems Studies. Examples of need for system
studies in the domestic heating area are these:
a. General guidance to the householder, indicating the
order of decreasing return on capital investment due to
different improvements in the heating system. This would
indicate the relative value of storm windows, attic insulation, wall insulation, frequent burner adjustment, etc.
A certain amount of such material is of course available
but it is inadequate. A serious study of how best to

present and disseminate such information is warranted.
b. Heat-pump economics. It is well known that the heat
pump gives a higher ratio of delivered heat to electrical
energy consumed in mild climates, and that in most areas
the winter heating load rather than the summer cooling
load determines the size of the unit. The use of heat
pumps is growing and, with their ability to deliver to
the heating space about 3 times the energy supplied electrically, should certainly be encouraged as a way of
making electric heating compare more favorably with gas
or oil. For 1 Btu liberated by the conbustion of fuel,
electric-resistance home-heating delivers about 1/3 Btu
into the home, fuel combustion in a home furnace about
2/3 Btu, and heat pumps about 1 Btu (varying as stated,
however, with mildness of climate). Oil-and gas-fired
absorption refrigeration units are of course also used
as heat pumps; they deliver to the heated space about 2
times the heat transferred by the flame to the evaporator,
or about 4/3 the heat of combustion of the fuel. Many
factors enter the analysis of heating and cooling costs
associated with mechanical and absorption heat-pumps.
Widespread dissemination of the results of heat-pump-
system studies of domestic heating and air conditioning
could accelerate their acceptance by the public. Heating
by this means is, relative to conventional heating, so
capital-intensive as perhaps to warrant a study of the
effect, on the national energy picture, of federal loans
on equipment purchases guaranteed to produce major reductions in energy consumption.
c. Oil vs gas vs electricity for house-heating. Each of
these energy sources has its advantages, some of which
carry no price tag. The householder who wishes to consider dollar cost, however, is in the cross-fire among three
purveyors of energy each making claims of superiority or
offering non-objective analyses of comparative costs.
Differences in price structure on equipment and installa-

tion and on fuel costs in different parts of the country
will make the positions of the three competitors shift
from area to area; the effect on equipment performance of
expenditures for maintenance may be a source of residual
error in the analyses, but something helpful to the public
can and should be presented.

d. Heat-storage applications. Section 2.4.4. presented
a table of heat storage materials indicating that, for
example, a cylinder of solid magnetic iron ore 2 feet in
diameter and 3 feet long would on being heated through
a 500 F interval, store 79 kilowatt hours (thermal) of
energy. Made into an externally insulated package with
flow-through passages for air, such a device could accept
electric power at night for next-day heating. An analysis
of whether the change in rate schedules which widespread
use of this device would justify would pay for its purchase is an example of storage applications warranting
study.

e. Total-energy systems. Homes need heat, power, and
refrigeration. A fuel-fired Otto-cycle engine operating
a generator set and supplying house heat via its exhaust
gas has been suggested. Combined with a heat-storage
device such a system might have economic merit. The
number of different combinations of elements and modes
of operation make a systems study of this field desirable.

f. Solar house heating. This has received attention in
Section 7.2. In competition with gas or oil, solar heating is today economically unattractive in the U.S.,
except in parts of the Southwest; Tybout and Löf (1970)
find, however, that it is competitive with electric
heating in most of the U.S. area if solar roof collectors can be bought for $4/sq ft. (Present prices, based
on hot-water heaters purchasable in Florida, Israel, and
Australia, are not much above that figure.) The increased
demand for domestic air conditioning offers a means of
spreading the fixed charge on the roof collectors over two

services. A study of solar systems for combined winter
heating, summer cooling, and hot water supply should be
made for enough locations to determine the optimum design
for each climate. There is certain need for auxiliary
heat in a minimum-cost solar system supplying hot water
and winter heating only; the effect of the more favorable
load factor of a combination system will probably be to
reduce but not eliminate the auxiliary heat supply.

Restriction of examples of systems studies to the
domestic heating and air conditoning area is not meant
to imply that there are not many examples available in
the commercial and industrial building area.

References

Agarwal, P.D., 1971. "Electricity Not Such a Clean Fuel,"
Automotive Engineering, pp. 38-39, February, 1971.

ASHRAE, 1969. American Society of Heating, Refrigerating,
and Air-conditioning Engineers' Handbook of Equipment,
p. 344.

Automobile Engineer, 1970. "Future U.S. Air Pollution
Regulations," pp. 145-148, April, 1970.

Broga, J., 1970. "Almost Pollution Free Powerplants are
Scheduled for Early 1975," Automobile Engineer, pp. 40-42,
December, 1970.

Chemical and Engineering News, 1971. Vol. 49, No. 27,
p. 35, July 5, 1971.

Heitbrink, E.H., and Tricklebank, S.B., 1970. "Electric
Storage Batteries for Vehicle Propulsion," ASME Paper
70-WA/Ener-7.

Heywood, J.B., 1971. "How Clean a Car?" Technology Review 73, No. 8, pp. 21-29, June, 1971.

NAPCA, 1970. Control Techniques for Carbon Monoxide,
Nitrogen Oxide, and Hydrocarbon Emissions from Mobile
Sources, NAPCA Publication No. AP-66, U.S. Dept. of Health,
Education and Welfare, National Air Pollution Control Ad-
ministration, Washington, D.C.

Tybout, R.A. and Löf, G.O.G., 1970. "Solar House Heating,"
Natural Resources J. 10, No. 2, pp. 268-326, April, 1970.

U.S. Department of Commerce, 1971. "Automotive Fuels and
Air Pollution," March, 1971.

Chapter 7

SPECIAL ENERGY CONVERSION SYSTEMS OF SECONDARY IMPORTANCE

7.1. Introduction

This chapter was to have covered fuel cells, thermionic and thermoelectric devices, and solar energy, as examples of energy conversion systems of secondary importance. Limitations on time prevented an adequate study of the literature on fuel cells, thermoelectric power and thermionic power. Fuel cells have been on the edge of practical significance for some years. The motivation for making them work has diminished over the years. In the 1920's the efficiency of power generation was not much more than 20% and fuel cells were spoken of as being capable of full conversion of the free energy of the fuel. Now the efficiency of power generation is 40% and the rosiest picture of fuel cell efficiency is 60%. But the prospects of commercial significance of fuel cells are good, and it is unfortunate that time ran out.

Thermionic and thermoelectric devices have significance as special-purpose devices, expensive and inefficient by central-power station standards, but unique for certain purposes.

Because one of the authors had many years of contact with solar energy research, time was found to write a section on that subject.

7.2. Solar Energy Utilization

7.2.1. Introduction and Precis

The enormity of the total energy reaching the earth from the sun has attracted countless individuals—covering the spectrum of quality from cranks to serious engineers and scientists—to the problem of considering the replacement of some of our present energy uses by solar energy. Some preliminary orientation is in order. The solar constant—the energy falling in unit time on a surface normal to a beam from the sun and external to the earth's atmosphere at the time-mean distance from earth

to sun--is 1.94 calories/cm^2 minute or 430 Btu/ft^{2}hr or
1.36 kW/m^2. From this one can calculate that the U.S.
annual energy consumption would be supplied by the solar
energy falling on the U.S. land area, if atmospheric
absorption were missing and the sun were directly overhead, in 1 hour and 56 minutes. If allowance is made for
atmospheric absorption and for earth geometry a more
meaningful limit is reached; based on the U.S. annual-
average solar incidence of about 1400 Btu/ft^2 day, the
continental U.S. intercepts about 600 times our 1970
energy consumption rate of 69x10^{15} Btu/yr. But this is
also misleading, because it assumes a conversion efficiency of 100%. With 25% of the U.S. energy consumption
converted to electricity at an average efficiency of 32%,
and on the assumption that solar energy is convertible to
process heat at 30% and to electrical energy at 5% (see
below for support of these numbers), one can readily
calculate that the 48-state total U.S. land area receives
as solar energy only 150 times our present energy needs.
A 1000 megawatt (24-hr average) power plant operating in
a 1400 Btu/day solar climate with an efficiency of 5%
would require 37* square miles of ground coverage, compared with a few hundred acres for a nuclear or fossil-
fuel plant. Major replacement of present power sources
by solar energy has poor prospects of success. The proposal that instead of covering the earth with solar
energy collectors we put up giant satellites to intercept
solar energy, convert it there to microwave radiation,

----

*
   In a recent radio broadcast June 21 the claim was made
that "an efficiency of power production from the sun of 10%
would yield 180 MW/square mile in the Southwest." One of
the best solar climates is El Paso, Texas, where the June
and January daily totals average 2692 and 1269 Btu/ft^2.
But except for 8 hours the intensities are too low to use,
and these numbers therefore drop to 2224 and 1240 Btu. At
10% efficiency of conversion, these numbers yield 24-hr
averages of 76 and 42 MW/mi^2, not 180; and 10% conversion
is twice what we have any realistic hope of achieving economically (see below).

beam it to the earth in a dilute enough beam not to make
a death ray of it, reconvert it to electrical energy and
then transmit it to our cities appears even less attractive in its long-range possibilities. It puts in series
at least four steps each one of which is far beyond our
present capability except at prohibitive cost.

Solar energy can be described almost completely by two
numbers, measuring quality and quantity. The quality of
sunlight is almost identical to radiation from a black
body or perfect radiator at 6000 K, which is a way of
saying that its thermodynamic potential or theoretical
maximum fractional convertibility into work is extremely
high. Said still another way, a high fraction of the
energy from the sun is in the form of shortwave radiation, capable of photosynthesis, of interaction with
atoms and electrons in crystal lattices (reference to
photovoltaic cells), or of coming to equilibrium with
high-temperature receivers through suitable wavelength-
selective filters.

The other number of the pair which characterize sunlight has already been given, the solar constant--
430 Btu/ft^{2}hr external to the earth's atmosphere. This
implies that solar energy is extremely dilute; its flux
density onto the earth is only one five-hundredth of that
onto the surfaces of a modern steam boiler. Unlike most
industrial process equipment, devices to intercept or
collect solar energy benefit little by scale increase.
Consequently, if solar energy is to find extensive use,
it will tend to be in small units to accomplish individually small tasks. Domestic hot water from the sun
is economically significant in many areas today, solar
house heating in some, and its prospects are improving;
solar distillation to produce fresh water from saline
water is economic in areas of extremely high fossil fuel
cost (certainly not in the U.S. mainland); solar electric
power from photovoltaic cells is significant in space

research where the laws of terrestrial economics are in-
applicable, and it has some chance of becoming much
cheaper. There are certainly enough of these areas to
justify a vigorous research program, but a major effect
on the national energy picture is not to be expected.[*]
A number of research areas will now be considered.

7.2.2. Flat Plate Collectors

If solar energy is to be used as a heat source, whether
for space heating, process heat or power production
through a heat engine, the collector can be a horizontal
flat plate or flat assembly of individual shaped collector elements, a flat plate tilted toward the equator a
fixed amount or a seasonally varying amount, an east-
west-axis cylindrical parabola rotated slightly with the
seasons, an equatorial-axis cylindrical parabola in
diurnal rotation, or a paraboloid on alt-azimuth or equa-
torial mount. Consideration of the progressive increase
in cost with progression along this sequence makes the
flat-plate collector assembly probably the most important
of the collection devices; and much research and engineering have gone into its design optimization. It is
characteristic of this device that it heats a stream of
air or liquid (usually water) the sensible energy in
which is then used at a remote point, and that the

----

[*] One of the authors was for 25 years Chairman of a University Committee charged with effectively spending the
income from a rather large solar energy endowment. During
that period solar energy research centers over the world
increased from three - at MIT in the U.S., in Algeria,
and in Tashkent in Russia - to some 35 others. The common pattern, with a few exceptions, was repetition of
older research and the addition of very little that was
new and less that was economically significant. To the
common plea of a developing country, "Our national resources are limited, but we have much sunlight and need
to learn how to exploit it as an energy source," the best
answer was "Only the affluent nations can afford solar
research for power production; its prospects are so poor.
But some uses of solar heat make economic sense."